ABOUT THE EDITOR

MALCOLM MACPHERSON is an author of eleven books of fiction and nonfiction, and a journalist whose articles have apeared in the *New York Times, Washington Post, Los Angeles Times, Playboy, Premiere,* and elsewhere. For years, he was a *Newsweek* staff correspondent, reporting breaking stories from around the world. His most recent book is *The Cowboy and His Elephant*. He lives with his family in Virginia.

ON A WING AND A PRAYER

INTERVIEWS WITH AIRLINE DISASTER SURVIVORS

EDITED BY

MALCOLM MACPHERSON

An Imprint of HarperCollins*Publishers*

HarperCollins books may be purchased for educational, business, or sales promotional use. For information please write: Special Markets Department, HarperCollins Publishers Inc., 10 East 53rd Street, New York, NY 10022.

FIRST EDITION

Designed by Philip Mazzone

Library of Congress Cataloging-in-Publication Data is available.

ISBN 0-06-095978-9

02 03 04 05 06 ❖/RRD 10 9 8 7 6 5 4 3 2 1

This book is dedicated to the men and women of the National Transportation Safety Board.

These narratives were edited from the National Transportation Safety Board's (NTSB) accident investigation reports, which are publicly available at the NTSB, L'Enfant Plaza, Washington, D.C.

CONTENTS

	Introduction	xi
1.	**Charlotte, North Carolina, July 2, 1994** USAir Flight 1016	1
2.	**Pine Bluff, Arkansas, April 29, 1993** Continental Express Flight 2733	17
3.	**Washington, D.C., January 13, 1982** Air Florida Flight 90	31
4.	**Cincinnati, Ohio, June 2, 1983** Air Canada Flight 797	49
5.	**Miami, Florida, May 5, 1983** Eastern Airlines Flight 855	63
6.	**Portland, Oregon, December 28, 1978** United Airlines Flight 173	71
7.	**Denver, Colorado, November 15, 1987** Continental Airlines Flight 1713	95
8.	**Flushing, New York, March 22, 1992** USAir Flight 405	107
9.	**Los Angeles, California, February 1, 1991** SkyWest Flight 5569 and USAir Flight 1493	125

10. **Honolulu, Hawaii, August 7, 1997** 137
Delta Airlines Flight 54

11. **Honolulu, Hawaii, February 24, 1989** 143
United Airlines Flight 811

12. **Detroit, Michigan, December 3, 1990** 155
Northwest Airlines Flight 299 and
Northwest Airlines Flight 1482

13. **Maui, Hawaii, April 28, 1988** 181
Aloha Airlines Flight 243

14. **Sioux City, Iowa, July 19, 1989** 191
United Airlines Flight 232

15. **Little Rock, Arkansas, June 1, 1999** 203
American Airlines Flight 1420

INTRODUCTION

What is it like to survive?

I do not mean survive another day of backstabbing on a deserted beach, like volunteers on the popular TV show. I mean, what's it like to *cheat death*, like those whom Fate chose to live through serious aircraft accidents? I sometimes carry this question like an anvil of dread onto flights and while I am listening to the attendant's safety briefing. I crane my neck for the sight of the nearest emergency exit door and doubt whether I am alone in a secret wish to hear from those who actually have lived to tell the tale. How did they do it, and who *were* these unwilling heroes?

In previous compilations—*The Black Box* and *The Black Box* (Revised Edition)—in which I provided Cockpit Voice Recorder (CVR) transcripts from the National Transportation Safety Board, I necessarily concentrated solely on what was happening in the cockpits. At the time I was editing those books, I regretted deeply not being able to show what the passengers were experiencing, too. I wished there were some device, like a CVR, to record passengers' thoughts and feelings, telling how they handled their fear and how they managed to extricate themselves through darkness, fire, smoke, and debris, once the wounded airplane came to a standstill. What did they do, consciously or unconsciously, to increase their survival odds? Or was getting out alive simply a matter of dumb luck?

Sadly, no such record existed. Or so I thought at the time.

Then, while on a visit to the NTSB's Washington, D.C., headquarters, I mentioned my regret to a charming woman named Nora

Marshall, who before joining the NTSB flew for an airline in an era when flight attendants were referred to as "stewardesses." Ms. Marshall pointed me to specific documents in the NTSB's "dockets" of crash investigations going back twenty years and more. I had never heard the name of the NTSB investigative group that she referred to (and that she leads)—the Survival Factors Group. But those earmarked docket pages gave me precisely what I had been looking for—a way to show what crash survivors endured, and best of all, told in their own words.

At the accident scenes, Ms. Marshall and her group, after being assembled and briefed, interview crash survivors, eyewitnesses and rescue workers. "There has always been a notion [at the NTSB] of finding out what the passengers were going through in an effort to prevent injuries in future accidents," Ms. Marshall explained. "We try to find out what happened by what the passengers tell us. They open up to us. Some people tell a little, and some a lot." The survivors often are prompted with written and spoken questions like these: "Do you recall the emergency instructions briefing received prior to or after takeoff?" "What actions did you take before and after impact to protect yourself and why?" "What if any are your final thoughts?"

Their survival narratives, when viewed collectively, have revealed opportunities for the NTSB to recommend improvements to the standards of air safety. They have prompted the installation of evacuation lighting strips on the floors of the fuselages, the manufacture of flame-retardant fabrics on seats, the improvement of cabin-crew emergency procedures and training, fire detection and suppression systems in lavatories and cargo compartments, heat-resistant slides to improve passenger escape paths, and, for instance, the design of seats that can withstand higher "g" loads of 16, up from 9—all factors in making survival more likely in a crash today.

And commercial airplane accidents *are* survivable, contrary to popular belief. The facts* may come as a surprise. *Ninety percent of all aircraft accidents worldwide are survivable*, according to the European Transport Safety Council (ETSC). Even more striking, the NTSB recently published their analysis: *For all accidents involving U.S. air carrier flights from 1983 through 2000, 95.7 percent of passengers survived.* And

*Not included is the toll from the hijackings and destruction of the four flights, United 175 and 93 and American 77 and 11, on September 11, 2001; those tragedies were not accidents, and they were not survivable.

in more than 46 percent of the most serious of these accidents—those involving fire, serious injury, and substantial or total damage to the aircraft—more than 80 percent of passengers survived.

"The most likely outcome for the serious survivable accidents is that most occupants survive," according to the NTSB's March 2001 report. "Although catastrophic accidents such as TWA flight 800 result in fatalities to all occupants, such accidents are the exception."

At the direction of Nora Marshall and with the further guidance of Ted Lopatkiewicz, the deputy director of Public Affairs, NTSB, I looked through the NTSB's docket list of "major investigations" for air incidents, keeping an eye out for those with the most responsive survivors. I sought to find a variety of scenarios, from airplanes that actually crashed to those, for instance, that slid off runways or blew a tire on takeoff rollout. Some interviews hardly contained a single reaction, while others were filled to exhaustion with detail and emotion.

Unwittingly, these responses often revealed a "voice" of the individual survivor that makes the narrative doubly interesting. We not only hear the details of a passenger's survival experience, we can intuit through their voices the outlines of who they are; each one is different, as you will see.

I have also chosen to include the narratives of some flight attendants who sometimes seemed to recall details and event sequences more vividly than passengers. In all these examples, I intentionally edited out the purely technical details, such as the minutiae of probable accident causes, pilot and cabin crew flying jargon, complete CVR and Air Traffic Control tape transcripts with time notations and technical addenda, interpretations of the flight data recorders, or findings of the NTSB's safety hearings. I set these elements aside and inserted words and phrases in brackets for ease of understanding. This book is intended for the general reading public; those aviation specialists in search of technical data would be better served to inquire directly from the government sources. My deletions are as much an attempt to help general readers reach an approximation of the state of ignorance of the passengers—who knew nothing of what the cockpit crews were doing or what was happening to the aircraft, beyond the obvious—as they are intended to streamline aviation jargon. Also, I chose to identify some aircraft crew members by name, but passengers, since their specific identities are irrelevant to the content of their statements, are denoted only by seat number, or by age, or sex.

So who does survive? Quite simply, the lucky majority. Passengers might even in a mysterious way forge their own luck by listening with care to the safety instructions of the cabin crews each time they fly, by identifying their nearest emergency exits and reading the safety-card instructions, by obeying the crew and responding out of "informed" instinct, and unlikely as this may seem, by possessing the *will* to survive. In the final analysis, as you will read in these pages—and to the surprise of no one—there is neither blueprint, seat number, section of the aircraft, time of day or airline to fly, nor magical recitation for survival. It is still just a roll of the dice, with the odds weighted heavily in your favor.

In this attempt to give expression to the survival experience in aircraft accidents, I hope the survivors' accounts you are about to read will inspire, intrigue, *and* inform. They are meant to leave or enforce the correct impression that airplane accidents *are* survivable, sometimes miraculously so.

ON A WING AND A PRAYER

1. CHARLOTTE, NORTH CAROLINA

July 2, 1994

USAir Flight 1016

USAir flight 1016, with fifty-two passengers, two cockpit crew, and three flight attendants, was executing a "missed approach" (when an aircraft has not landed normally and is going around to make another approach to land) while attempting to land on runway 18R at Charlotte, North Carolina's, Douglas International Airport at 6:42 P.M. in "instrument meteorological conditions." It was raining hard. "Convective activity that was conducive to a microburst [wind shear]," according to the NTSB's accident report, punched the Douglas DC-9-31 to the ground. It collided with trees, breaking up, skidding like a toboggan down a residential lane, and smashed into a private residence, catching fire. The captain and one flight attendant suffered minor injuries. The first officer, two flight attendants, and fifteen passengers sustained serious injuries. The remaining thirty-seven passengers died. Impact forces and a post-crash fire destroyed the airplane. No one on the ground was injured.

COCKPIT VOICE RECORDER KEY

US1016:	Radio transmission from the flight
CAPT:	Captain cockpit communication
FO:	First officer cockpit communication
APPR:	Charlotte Approach Control
TWR:	Charlotte Tower

Despite a thunderstorm "cell" advancing on the area, the cockpit crew had every reason to believe that a smooth approach and landing were possible.

APPR: Tell you what, USAir 1016, [you] may get some rain just south of the field. Might be a little bit just coming off the north. Just expect the ILS [instrument landing system] now. Amend your altitude . . . maintain three thousand.

CAPT: If we have to bail out [go around on the landing]. . . [*Pause*] It looks like we bail out to the right.

FO: Amen.

CAPT: Ten miles to the VOR [navigational aid] which is off the end of the runway. 'Bout a mile off the end of the runway.

FO: Yeah.

CAPT: So I think we'll be all right.

CAPT: Chance of [wind] shear.

FEMALE, AGE TWENTY-EIGHT [WITH HER NINETEEN-MONTH-OLD DAUGHTER], SEAT 19F

[During the flight] my daughter moved back and forth between her seat and my lap. She wanted to play with the people in the row behind her. She got tired and laid down next to me. Her head was in my lap. After the flight attendant collected the lemonade cups, I felt a little bump and then a big bump, and the airplane just dropped. I could not understand what happened. The weather had been sunny, and I had seen thick white clouds. I heard an announcement, "I'll have us on the ground in about ten minutes," and, "Flight attendants, please prepare for landing." I recall entering rainy weather, and I leaned forward in my seat to look out the window. The rain was coming on the wing slanted.

COCKPIT VOICE RECORDER

Other traffic in the sky near the airport and on the field's taxiways was aware of the storm, talking about it with the ground controllers and the tower.

TWR: And [USAir] 806, looks like we've gotten a storm right on top of the field here.

US806 [on the ground, waiting to depart from Charlotte]: USAir 806, affirmative. We'll just delay for a while.

TWR [to USAir 1016]: Charlotte Tower, runway 18 Right, cleared to land. Following an FK[okker] 100 on short final. Previous arrival reported a smooth ride all the way down final.

US1016: USAir 1016, I'd appreciate a pirep [pilot flight report] from the guy in front of us.

FO: Yep, [the storm is] laying right there this side of the airport, isn't it?

CAPT: Well.

FO: The edge of the rain is, I'd say. . . .

CAPT: Yeah.

TWR: USAir 1016, company FK 100 just exited the runway, sir; he said smooth ride.

TWR: USAir 1016, wind is showing 100 at 19.

FO: One hundred at 19, eh?

TWR: USAir 1016, wind now 110 at 21.

CAPT: Stay heads up.

TWR: Wind-shear alert, northeast boundary winds 190 at 13. Carolina 5211, Charlotte Tower, runway 18R, cleared to land, wind 100 at 20. Wind-shear alert, northeast boundary wind 190 at 17. USAir 806, you want to just sit tight for a minute, sir?

US806: Yes, sir, we'd like to just sit tight.

TWR: USAir 797, company aircraft in front of you is going to sit and wait a while, sir. Do you want to go in front of him?

US797: No, no, it wouldn't sound like a good plan. We'll, uh . . . It didn't look like a whole lot [of rain] to us on the radar taxiing out, so it shouldn't be, uh, shouldn't be too many minutes.

CAPT [USAir 1016]: Here comes the wipers.

FO: All right.

Start of sounds similar to rain and sounds similar to windshield wipers.

FEMALE, AGE FORTY-FOUR, SEAT 19D

I felt the engines, brakes or whatever, and then I felt a spinning like turning the engines to go, like [the captain] was circling. I was sitting next to the engine. I heard slowing down of the engine, and then the gunning up on the engine.

FEMALE, AGE TWENTY [WITH A NINE-MONTH-OLD INFANT], SEAT 21C

The flight attendant kept peeking around the corner and smiling and making my baby laugh. I was seated in an aisle seat in the last row in front of the flight attendant. There was a little boy on the left side of the airplane at the window in front of me, and no one else was to my right. We were the only occupants of the row.

I heard the pilot make an announcement for landing and say that it was "ninety-something" degrees, and sunny out.

COCKPIT VOICE RECORDER

FO: There's, oh, ten knots right there.
CAPT: OK, you're plus twenty. Take it around; go to the right.

At this point, the captain orders the first officer to abort the landing. On a missed approach they choose to go around to try again and as such they are "on the go." The aircraft is about two hundred feet off the ground.

US1016: USAir 1016's on the go.
CAPT: Max power.
FO: Yeah, max power. . . .
TWR: USAir 1016, understand you're on the go, sir. Fly runway heading. Climb and maintain three thousand.
FO: Flaps to fifteen [degrees].

FEMALE, AGE TWENTY-EIGHT [WITH HER NINETEEN-MONTH-OLD DAUGHTER], SEAT 19F

My seat belt was pulled so that it was fitting me. When I realized something was wrong, I leaned forward and pulled my daughter toward me. [She repeatedly demonstrated to a Survival Factors Group interviewer a motion of leaning forward and wrapping her left arm over the child.] I thought I could use both arms to grasp my daughter as I leaned over.

MALE, AGE TWENTY, SEAT 21D

Then suddenly we were in the middle of a dark rain cloud. As soon as we entered the dark cloud, the pilot goosed the gas and sped up. Then I felt the plane bouncing and then a jolt. In a split second everything opened up.

COCKPIT VOICE RECORDER

CAPT: Down, push it down.
US1016: Up to three, we're takin' a right turn here.
TWR: USAir 1016, understand you're turning right?

The flight's altitude begins decreasing below 350 feet. A second later, a "whoop whoop terrain" sound begins and continues until the first sound of impact.

CAPT: Power. . . .

Sounds of impact.

FLIGHT ATTENDANT RICHARD DeMARY*

Probably the first real sign of trouble . . . was the sinking sensation, knowing that this is just not right . . . hearing—*knowing* we were off the airport—and hearing "Terrain! Terrain! Terrain!!" [Recorded warning from the cockpit]

Hearing that, and then [followed] almost immediately [by] the impact and fortunately or unfortunately, you know, being conscious

*A copy of DeMary's interview was published in the March 1995 issue of *Cabin Crew Safety*, a publication of the Flight Safety Foundation.

through the whole thing, the whole crashing process began. It happened so fast. Initially it was disbelief and then just the terrifying feeling that we're crashing.

My recollection was that there were two impacts. You know, some people say there were three, but I remember two. The first impact with the ground, the sound of trees breaking, at that point knowing we were crashing—just the force of the impact was extremely violent, almost takes your breath away when you're crashing like that. Then immediately after the first impact, the second [and] most violent impact, to me, is when I think we hit a tree. The airplane hit a tree, basically peeled back [that] one side of the airplane—broke the airplane apart into three sections. The nose section, with a few passenger seats, went off to the left.

I was in that section, the nose section. One part of the airplane—and I believe it was the center part of the airplane—from the first-class seats back to just past the emergency exit rows, I believe—I don't know if this is proper to say—but it basically wrapped around a tree because that's what happened. We hit a tree and [the aircraft broke into separate sections] and then the tail section proceeded to go into [the carport of a house].

FEMALE, AGE TWENTY-EIGHT [WITH HER NINETEEN-MONTH-OLD DAUGHTER], SEAT 19F

After the crash, I remember pulling on my baby's leg to get to her.

I was in and out of consciousness and did not remember how I got out of the airplane. It was difficult for me to open my eyes fully. When I came to, I saw blood on my daughter's face. I had dreamlike memories of hearing a woman asking, "Where is my baby? Where is my baby?"

FEMALE, AGE FORTY-FOUR, SEAT 19D

I recall a violent shaking and a big burst of fire made me think that I was going to die. There may have been up to five jolts. I was thrown up, back, over, back, and then the airplane came to a stop.

Immediately after the impacts, flames blew up into my face and singed my hair and the parts of my body that were not covered by my clothing. There was fire in the airplane when it came to a dead stop. Then the fire receded.

I looked around and saw that I was the only one in that part of the cabin. It was empty, dark, and totally quiet. Then I began to hear people saying, "Please help me. Please help me. Please help me out. I can't find my baby. I can't find my baby."

FLIGHT ATTENDANT RICHARD DEMARY

I remember just after we hit the tree, the feeling of the rain hitting us, the wind, the noise, because we were in a darkened, enclosed cabin as we were coming in to land and then all of a sudden we were opened up to all of the elements and I remember feeling the rain hit me and the noise—screeching of metal, hearing the rain hitting the airplane, feeling it hit me—and it happened so fast that I guess I've never really had the opportunity to sit down and think about all the different sounds, but just noise. Probably the loudness came from the scratching of the airplane on the ground and on the pavement because the section I was in slid down a street.

I don't remember hearing people. I don't remember that now. Whether or not I ever will, I don't know.

I don't remember the wind, but I do remember the rain and I remember having jet fuel on me. Where it came from I don't know, maybe from puddles of jet fuel around or a spray of jet fuel or something. I'm not sure. But really, nothing came to mind immediately other than [that] I couldn't open my door and there was no reason to open my door. I was in the open.

There was no cabin there. *There was no more airplane.* There were a few rows of seats on the nose section of the airplane that I was in, but the airplane had actually broken apart and bent.

I remember . . . sitting in my jump seat, not seeing anything once the airplane came to a stop, and at that point knowing we were in a crash, knowing that it's time to get out, it's time to evacuate, and I immediately went for my seat belt [and] started yelling, "Release seat belts and get out! Release seat belts and get out!" which is the first command that we would yell upon coming to a stop, and at that point

Shelly [Markwith, a flight attendant sitting in the jump seat beside DeMary] . . . was yelling her commands, "Release seat belts and get out!" and trying to get her seat belt unbuckled. Shelly told me, "I can't get out. My legs are broken. I can't get out."

I was actually leaning on Shelly. I stood up and I had to kick my feet free from debris. As I stood up, I believe I saw the captain crawl out of the cockpit, through the cockpit door. Because of Shelly's injury she couldn't do anything. She had shattered her kneecap and I think she had a cut to her bone on her thigh, and she had some burns and she—she couldn't even crawl.

Shelly was having some difficulty getting herself unbuckled, so I unbuckled her seat belt, bear-hugged her, grabbed her and picked her up. She couldn't stand because her leg was severely injured and she really couldn't do anything to evacuate herself. So I grabbed her and just carried her and stepped off to the street. Five feet away from the airplane she fell again, and then I grabbed her wrists and just dragged her away, just trying to get her to a safe distance away from the airplane.

FEMALE, AGE TWENTY [WITH A NINE-MONTH-OLD INFANT], SEAT 21C

Then there was a bump, and my baby was crying as she flew out of my arms. I tried to hold on to her. There was a bump. There were a lot of bumps. The wheels touched the ground, and we were bouncing all over the place, and I hit my head on something. People were flying all over the place.

MALE, AGE TWENTY, SEAT 21D

I felt stuff, dirt, hitting me in the face. The next thing I remembered was laying on my back under a pile of metal with everything piled on top of me. I barely had enough room on my chest to exhale and inhale. My lung had collapsed.

I felt like I was in the wreckage for about an hour before I was pulled out. I was on my back, facing upward. I could only see metal and a little bit of light. I felt a lot of heat on my face. I could hear a few people screaming and asking for help. After a few minutes of lying in the wreckage I heard someone say, "Is anyone in there?" Four or five people yelled out, and I yelled out, too. It was so tight in there that I really couldn't yell that loud because I couldn't get enough air.

FLIGHT ATTENDANT RICHARD DEMARY

Then the awareness of being in the accident, that now, I have survived, that I have to do something, came full force. It became very important for me to help anybody I could help, not only to help but to search, to find people, and I couldn't wait for them to come to me. I had to go find somebody. I was completely disoriented as far as where the rest of the airplane was. At that point, the thought crossed my mind that we were the only ones who survived.

[The] fires were bad enough that I could feel the heat, so I knew that we had to get away or the people [who] were too injured to do anything [to escape] had to get away from [the heat]. It was—it was hot! A lot of fires were spotted around the area, a lot of small fires, and then I remember the smoke and feeling the flames and seeing the flames over quite a large area which turned out . . . to be by the tail cone, the back of the airplane. There was a lot of fire. [The nose section] broke off and came to a stop in the street just to the left of a house—in front of the house. I helped Shelly away into a grassy little area of yard and I was confident that she was safe at that point. But after helping Shelly . . . I had no airplane. There was just nothing there.

I [suddenly realized] that, "Oh, my God, this is a residential neighborhood!" Because I saw the houses. I saw the trees. I saw the street, the sidewalk, and I think I immediately thought, "What are we doing here? This is not right!" Because I always thought, "Well, it's reality that we might crash sometime," but I never thought it would be into a house or into a residential neighborhood. I mean we were in *somebody's yard.*

FEMALE, AGE TWENTY-EIGHT [WITH HER NINETEEN-MONTH-OLD DAUGHTER], SEAT 19F

I awoke outside the airplane where someone had dragged me away from the wreckage. When I awoke my head was in the lap of a man. . . . My daughter was being held by a woman. . . . I heard my daughter crying, and I told her, "Mom's here."

FEMALE, AGE FORTY-FOUR, SEAT 19D

I believe that instinct kicked in when I realized there was smoke. I tried not to breathe any smoke by taking only shallow breaths. I told myself, "Don't panic! You can get out!" I kept trying to focus on, and believe in, my ability to get out. I believed that my positive thinking helped me to survive. I did not panic. I did not want to die, and I intended to do everything necessary to prevent it. I decided that if I was going to die, then it was God's will.

I unfastened my seat belt and stood up. It's too incredible to explain the position of all the seats. The seats were down under [upside down?]. I stood up and looked around to position myself. From my back right side I saw the flight attendant crawl over a bunch of stuff. I asked the flight attendant, "What can I do to help?" She replied, "Come try to help me open this back rear emergency door." About the same time a black man who had been sitting to my left escaped from under his seat. He crawled up through the wreckage and also came back to help the flight attendant.

FLIGHT ATTENDANT RICHARD DeMARY

At that point I took off my tie. I don't have a memory of little bits of what I did. It's my understanding there was a lot of fire and possibly a lot of bodies, you know, my mind just doesn't want me to have it right now. But I remember ending up by the tail section and it was very quiet. I didn't hear anybody, didn't see anybody. There was a break in the airplane, a break in the fuselage, and at that point, I thought, "Well, I have to do something!" and I started yelling my commands. I thought, "Well, it's a starting point. If [people are] in shock, if they hear, 'Release seat belts and get out!' it's going to give them the starting point!" So I started yelling, "Release seat belts and get out! Release seat belts and get out!" I'm continuously yelling it, as I'm walking, as I'm looking for somebody, looking. . . .

I didn't actually go [back] in the aircraft. I was right beside it, right next to the engine. There was just a small break in the right side of the fuselage. That side was fairly intact.

I had really given up at one point. I thought, "Well, there's probably nobody that survived—that survived the impact," but I remained. I continued to have faith that somebody might have survived—you know, somebody might be in there. I remember how hot it was. The fire was tremendously hot.

Then a woman appeared at that break [in the fuselage] with a baby. She was able to get out of her seat belt and this was probably some time after the accident [before any rescue squads had arrived on the scene]. . . . It seemed like an eternity but she came toward my voice.

. . . She appeared at that small opening [in the fuselage] and I reached in and I grabbed the baby . . . and grabbed her arm and pulled her. I mean, it wasn't like Shelly. I literally [had] dragged Shelly on the ground, but I'm sure I just grabbed [this woman with the baby]. She was yelling, "Help me! Help me!" when she was in the airplane.

There was a small shed in the backyard [behind the house carport, which the piece of aircraft fuselage had struck] and I just took them back there to safety.

I went back to the airplane again, and I remember thinking how hot it was—I mean, I placed my arm on the engine [cowling] and it just burned all the skin off my—not all the skin but it severely burned my arm and just the heat of it, the heat of the metal, and I remember hearing explosions, small explosions, and I thought, "Well, I have to do what I have to do, but I can't stay here forever." I was concerned that I would succumb to the smoke or, you know, the fire or something like that.

But anyway, I did go back, and I continued to yell, "Release seat belts and get out!" and another woman appeared at the same opening. . . . She was yelling, "I don't want to die! Help me! I don't want to die! I can't find my baby!" I later learned that she was one of the women who had a child, a lap child [i.e., a child without a seat or seat belt of its own] with her.

There were two lap children aboard the flight. One survived, one died.

I helped her out of the airplane and she had some injuries, I think, because she was basically immobile. It took a lot to get her out. The tail of the airplane was on the ground, but the center section was in the air. It was at quite an angle. So the opening was probably . . . mid-waist to chest-high.

I had to reach up just a little. She's yelling, "I don't want to die! I don't want to die! Help me out! I can't find my baby! I can't find my baby!" I literally had to just bear-hug her and pull her out because she was heavy. Anyway, I got her out and got her back to the same place I took the others at the back of the yard.

You know, she was yelling, "I can't find my baby!" and I went back then, but at that point I didn't think it was probably . . . [the] best thing to do. What I'm saying is, I didn't think it was appropriate that I actually go into the airplane and search, because of the fire and the smoke and how long it took them to get out.

But I [started] back again after helping the lady out and [I saw a man and a woman] from the neighborhood. I asked them just to stay with the passengers [in the backyard] and I went back to the airplane and continued to yell. . . . There was nobody. And at that point, well . . . I thought, "I need to get away because it's very hot and I don't want to survive the impact to die in the fire of the secondary explosions." Something like that. And I thought I could be of help somewhere else, possibly.

And one of the things that bothered me, too, is that I did have jet fuel on me. My clothing was flammable and probably more so with jet fuel on me. So I went back around the back side of the house, toward the front, and I saw the captain, and at that point Fire and Rescue still hadn't arrived. I remember hearing the captain say, "She's OK! She's OK." I thought he was talking about Shelly, but in reality he was talking about Karen [Forcht, the third flight attendant on the airplane], and then I did see Karen, and she had severe burns. She had lost her shoes in the impact and she had severe burns on her arms, hands, face, legs. I believe three people had followed Karen out when she got out.

FEMALE, AGE FORTY-FOUR, SEAT 19D

We pulled the door open a little, and flames were visible. The flight attendant said, "No. No." And we shut the door immediately. Then the flight attendant turned around and said, "We can't get out this way." And we went the other way.

By the time I refocused and turned around there was nobody else there. I saw the [same] black man and a small black child wiggling out of a place where there was light at the tail end of the airplane. I had previously seen this area of light when I went to help with the door, but I ignored it because I had seen flames there as well as light.

Just as Flight Attendant DeMary had trouble believing that the airplane had come to a halt in a neighborhood with trees and houses, this passenger was equally baffled, and reasonably so, when she looked down before stepping out of the wreckage and saw a household's kitchen in front of her.

I came back to where my seat was, and I noticed that to the left there was a door going into the kitchen [of the house that the aircraft had collided with]. I was confused. I asked myself, "Why is this kitchen here?" I tried to open the kitchen door, thinking that I could get out through the kitchen. The door was like a storm door with glass on the top and a white bottom. The glass in the door was not broken. I wanted to smash the window, but I could not find anything loose that I could use.

I decided that I [had to escape] through the area that I had seen the man and his young son wiggling through. Then I heard a man yell, "We made it! We made it!" And I knew I had to go that way.

FEMALE, AGE TWENTY [WITH A NINE-MONTH-OLD INFANT], SEAT 21C

During the crash and the impact of the airplane hitting the ground, my baby went flying in front of me. I tried to hold her and I couldn't. They had told me I could hold her in my lap. I would have paid for her to sit in a seat. They said she did not need a seat. The man said that she did not need a seat because she was under the age of two, and that she was a "lap baby," and I could hold her. I would have given my life for her. She wasn't here that long. She had just turned nine months old.

MALE, AGE TWENTY, SEAT 21D

I could hear people above me. It sounded like people [were] walking on metal and banging metal around. I heard someone say, "Where are you?" And all I could say was, "Here!" because I did not know where I was. Someone put an air mask on top of my face. I could hear cutting, and when they used the cutters I could feel vibrating around me. They finally got to me and cut me out.

FEMALE, AGE FORTY-FOUR, SEAT 19D

My seatmate [in seat 19F], whom I had not seen before, now came to and began yelling, "Please help me, please, someone help me." He had a tree lying on top of him, and I could only see his head and one boot. I asked him, "What do you want me to do?" I could not get him out. I told him to try to think of something else. I knew I could get out, and I would send some help.

I did not think the [fireball] caused my burns. I think that I got my burns when I shimmied out of the airplane. Everything that was metal was so hot it was like touching a hot griddle. As I got out through the tail I was still about eight feet off the ground. I saw the hood of a car that was slanted, and I slid off of it and ended up at the front porch of the house that was connected to the carport.

FEMALE, AGE TWENTY [WITH A NINE-MONTH-OLD INFANT, WHO HAD DIED], SEAT 21C

I heard my baby calling to me after the airplane came to a rest. She was calling me—"Mama." A man . . . pulled me out of the plane and told me that my baby was OK. They said she was OK and that she was at another hospital.

The woman was unable to restrain her nine-month-old child during the impact and her child was thrown forward over three rows of seats and killed by a massive trauma.

FLIGHT ATTENDANT RICHARD DEMARY

At that point a few other passengers were coming out of the wreckage. Usually, when you think about an accident, you think everybody's going to be going through the same thing. Karen's [experience] of the accident was a little bit different because she was in the back of the airplane that broke apart. She had the impact. She had the debris flying through the cabin. She had the fireball. She had the smoke. So she had a lot of different elements to contend with. And then probably most importantly, Karen had the element of not having a usable exit. She was able with the assistance I think of a couple of passengers to get that back [emergency] door open [to] access the tail cone and found it unusable, [and] immediately closed [the door]. That's where the fire was. There was smoke in there.

There was a triage area forming, basically, to the left of the house. So there was no sense in me staying with the people [who] were, I guess, OK. I went back toward the front of the house and I remember seeing a kid from the neighborhood—this thirteen- or fourteen-year-old kid—and I said, "Is anybody at home? Is anybody in the house?" and he said he didn't know, so I thought, "OK, the next thing to do is to go into the house," because the airplane was too hot, there was too much fire and I wasn't going to go inside the airplane.

I'm talking to that young kid in front of the house and I thought, "Well, if anybody's home, we need to see about them." So I went to the front door and just as I was running to the front door, a passenger crawled out of the wreckage who said, "Somebody's in the garage."

I thought, "Well, there are people home. There is somebody there who needs some help!" There were cars within the wreckage, from the driveway, so I went to open up the front door and I thought, because it was locked, there's probably nobody home. So I kicked in the front door and just looked to the left a little bit and the captain followed me in and I think the young boy from the neighborhood followed me in. There didn't appear to be any damage to the left inside of the house.

Then I looked right and I saw the living area or the dining room area with a set table. I remember seeing the mats on the table, and then I looked and I saw the door that [opened onto] the garage. It was actually just a carport, and it opened to the inside and then there was a [storm] door, and [the door] opened to the outside. I couldn't open up that door because of the debris within the garage area and—so then I just busted out the glass of the door.

And then I heard a voice [in the carport], and then I started just speaking with the guy, yelling to him. He was yelling, "Help me! I can't breathe!" and I was yelling back to him, "Cover your mouth if you have anything to cover your mouth with, and breathe through that!" and he was yelling back, "I don't have anything to cover my mouth with." I couldn't see him because the smoke was very heavy.

It was kind of—it was a grayish smell of like, plastic burning, just very heavy. I couldn't even breathe it in, and I was getting good air [from inside the house] with that bad air [in the carport], and I found it difficult to breathe.

There was so much debris, and I remember one of the main-wheel tires was standing right there next to the house, just within the debris, and I couldn't see the guy. I couldn't really make out anything in that area, and I yelled for him to cover his mouth if he had anything to cover his mouth with, and he said he didn't, and then I shouted for him to stay calm, try to relax, breathe slowly, just to stay calm, that help was on its way. At that point I heard the fire trucks arriving. I said [to the man], "Somebody's here to help."

I ran out to tell [rescuers] that somebody was in that carport area. The fire trucks couldn't get in because we did crash in a residential neighborhood, and [the aircraft had] sheared some telephone poles. [Another guy and I] moved the telephone poles so that [the fire trucks] could get out, or get back in to a closer area, and I told them that there was somebody in [the carport]. The gentleman did survive. He was a passenger.

So the fire and rescue trucks were arriving. [I] helped [the firefighters] pull out some hose [from a fire truck] and then was asked to get away. I was told, well, you know, "Your job's done. Just get away."

2. *PINE BLUFF, ARKANSAS*

April 29, 1993

Continental Express Flight 2733

Continental Airlines' Continental Express flight 2733, referred to as "JetLink," was passing through 17,400 feet in its climb out of Little Rock, Arkansas, when without warning the twenty-seven passengers felt severe vibrations throughout the airframe. The airplane spun left and plummeted 5,500 feet before the first officer lowered the landing gear to stabilize the airplane. Control was regained. When the passengers looked out their windows, they could see that three (of four) propeller blades of their Brazilian Embraer EMB-120RT's two engines had disintegrated, along with the engines' upper cowlings. Maintaining level flight was impossible, and the captain, who was in control of the airplane, headed for the nearest airport at Pine Bluff, Arkansas. He had trouble turning the airplane to the right, and he overshot the runway, touching down with only 1,800 feet of rain-washed landing surface to spare. The airplane hydroplaned out of control, skirting a truck and a person standing out in the open but hitting a three-foot-high gravel embankment, tearing off the right wing, before rolling off the end of the runway and coming to a stop in mud.

FLIGHT ATTENDANT AIMEE YATES

The takeoff was normal for such a heavy load. I was looking at the passenger list. The two girls in the first row were both named Heidi and I noticed we had four Gold Elite Members on board and several Silver and One Pass Members. The gentleman in the second row on the right was a Gold Member and so was the man in the first seat on the left. I was trying to decide if I would serve complimentary champagne that we were celebrating our "coming out of bankruptcy" with. The champagne bottles took up all my trash space. With a full flight and a regular beverage service, there would be a lot of trash. I would have to use the lavatory trash container for the soda cans.

After deciding I would serve the champagne I got out of my jump seat and stuck my head into the cockpit. I told Captain [Ray] Granda that I had decided to serve. I walked down the aisle to prepare the galley cart. We were still climbing. I pulled out the cart, set the brake, opened the door, set the glasses up, and got out the champagne to serve. I left the champagne bottle and the cups lying down so that when I brought out the cart they wouldn't fall off.

I walked "uphill" to the cockpit to see what altitude we were climbing to, where we were, and how much longer until we reached cruising altitude. The altimeter indicated twenty-two thousand feet. We were at sixteen thousand feet when I reached the cockpit. Captain Granda was paying attention to the instruments and the airplane was on autopilot. Frank [First Officer Frank Chenevert] was writing in the log book and eating his crew meal. I asked the captain how much longer till we reached altitude. He replied, "We're almost there." There was some conversation between Frank and I, and [Captain] Ray [Granda] and I. I was waiting until we leveled off before I rolled out the cart.

FEMALE, AGE FORTY-TWO, FRONT AISLE SEAT

When we first entered the plane [my friend] Heidi was moving toward the window seat in the first row. I sat in the aisle seat in the first row next to her. I carried on a bag because it was filled with a lot of my glass souvenirs from my vacation, and I did not want them to break. The stewardess told me there was no room for the bag on board and roughly handed it to the man who put it in the luggage compartment. [The same stewardess] put another bag of someone else between the pilot and copilot's seat on the floor and [she] told me to put my feet on it if it started to fall back on takeoff. I couldn't even reach it with my feet from where I was sitting. I asked her how long the flight would take, and she told me from one and a half [hours] to two hours . . . knowing the pilot. She was talking to the pilot and copilot, and it sounded to me like she was arguing with them and was mad. I mentioned it to Heidi that I didn't like [the flight attendant's] attitude, because she sounded mad, but then she started laughing, so I guess everything was OK. She was eating carrot sticks [that] the copilot gave her. Then she came and did her demonstration and had me hold on to the mask while going through the routine. All the time I was watching the pilot and copilot, telling Heidi I was learning how to fly the airplane, since the [cockpit] doors were open all the time. I watched the pilot turning a wheel which I think made us turn to the right. He kept turning a knob. It looked like for balance. I was watching the little screen above them with a map on it and telling Heidi, "That's how they fly a plane. They stay on course by staying on line with the middle line running up and down on the map." I told her we were flying first class sitting here and we would be served first. I kept telling her I didn't like this airplane, because it was so loud, and I don't like flying on small planes. [Heidi] kept telling me to quit panicking and to be calm. I asked the stewardess for something to drink, because my throat was dry. She said not until we reached altitude, and she went to the back of the plane to check on something.

At that time the pilot took off his seat belt, pushed his seat back and put his right foot up on something between the seats. He turned to the copilot and was talking to him. I told Heidi I didn't like him not looking where he was going. She again told me to calm down.

COCKPIT VOICE RECORDER

While the aircraft is still climbing, the flight attendant stands in the cockpit door and jokes with the captain about an insect that is stuck on the windshield.

FA: Oh, do the windshield wipers. That'll wipe it [the insect] off.
CAPT: Naw. I can't. Naw. I can't do it 'cause we're goin' too fast.
FA: That'd be funny. That'd be funny. That'd crack me up. That'll make my day.
CAPT: . . . Off.
FA: Do it, do it, do it.
CAPT: Okay.
FA: Can't we climb any faster?

CAPT: Why do you want to get up so fast?
FA: [Because] I can't pull [the service cart] uphill.
CAPT: Okay.
FA: The last time I had to pull it up the hill.
CAPT: We'll try to get up a little more.
FA: We'll try to get up there?
CAPT: Yeah. We're almost there. Another six thousand feet, another six minutes.

Here the FA jokes with the captain, offering him something that the CVR tape does not make clear.

FA: Gimme, gimme, gimme. Climb. Gimme, gimme, gimme.
CAPT: You're right. Better to receive than to give.
FA: What?
CAPT: Ah, like you said. Better to receive than to give.
FA: I don't know.
CAPT: I live by that.

The captain and the flight attendant engage in another five seconds of what the NTSB characterized as "nonpertinent" conversation.

CAPT: We're not climbin' very fast. Yeah, we're really heavy.

At this point, ice is starting to form on the wings, and the inputs in the autopilot for vertical mode are creating an instability in the aircraft. Its "heaviness" is an indication that it is about to stall.

FA: I know. . . . I know. [I couldn't, or you couldn't] walk, like walk a straight line [and] not sway all over it.
CAPT: [laughs]
FA: So that's what you meant by a weight problem?
CAPT [to first officer]: Frank, hang on. This ain't right.

Two seconds pass. The stall-indicating stick shaker [the control yoke in the cockpit that shakes violently to warn the crew of a stall or of a sudden loss of altitude] starts to shake the control yoke, and the autopilot automatically disconnects. The airplane plummets. The flight attendant is thrown back into the passenger cabin.

CAPT: Airspeed. . . . Hang on. Hang on. Power's up. . . . Power's . . . up. I think it's an overspeed. It's an overspeed.

FLIGHT ATTENDANT AIMEE YATES

When we reached eighteen thousand feet the aircraft began to yaw right, then left, and shake. It felt like turbulence. Then the vibration grew stronger and stronger very quickly. While I was still in the door, Ray said, "Who touched the trim? Frank, did you touch the trim?" I remember Ray saying, "It's all the way to the right." As the vibrations got stronger it got harder to stand up. Frank shouted for me to take my seat. The vibrations were growing and the pilots were trying to figure out why and what was wrong. They were reciting a checklist.

I saw the slip indicator [on the instrument panel in the cockpit] with the bubble far left, and the altitude indicator went from blue and brown to mostly brown. I was thrown back, past the second row. The shaking was very violent and the window shades were closing. The stick shaker went off. I heard a sputtering sound from the left of the airplane. And it sounded like things were falling off [the airplane]. It felt like we were going to shake apart. There was no screaming in the airplane.

I remember climbing to my seat, I'm not sure how. But I managed to get into my jump seat and strap myself in as tight as I could. The vibrations continued and I watched as things began to spill out of the cart. Almost all the shades were closed. I could hear Ray and Frank talking about what went wrong with the left engine. Then the pitch of the aircraft changed. We rolled left and headed straight down, nose first into a spin. I heard Frank scream, "Ray, we're in a spin!"

The spin was very violent. The passengers were pressed into their seats and their faces were distorted from the g forces and the spinning. The two girls in the front had been leaning over with their heads down and holding each other when the vibrations started, but everyone was pressed into their seats.

At this point I could hear Ray shouting, "I can't believe we are going to die, Frank! Frank, man, we're gonna fuckin' die!"

Frank was still reciting a checklist as we were spinning. We were losing altitude rapidly, and Frank was still working to get us out of the spin. Apparently something had happened to the left engine or prop. With the right engine running it aggravated the spin and increased our descent. Then it got very quiet. There was no engine noise. Just air passing by as we dropped.

At this point I had given no instructions to the passengers. When it became quiet, I had accepted that we were going to die. It seemed that everything Frank and Ray were trying wasn't working, and now we were just going to fall out of the sky. It seemed like forever we dropped. Then Frank dropped the [landing] gear and we stopped spinning and began to fly again. I heard the gear go down. I never doubted the ability of the pilots until the engines quit. I knew they could fix it.

FEMALE, AGE FORTY-TWO, FRONT AISLE SEAT

Then the plane shook really bad, and I noticed the look on the pilot's face like he was scared to death.

I told Heidi I knew something was really wrong now by the look on his face. I felt the plane jerk to the right. The stewardess came back up and put her arms on both the pilot's and copilot's shoulders and immediately returned to her seat. I put my hand down in Heidi's lap, because of the plane shaking, and she tried to calm me down.

The stewardess then told us to put our heads down in crash-landing position. I could hear the pilot yelling cusswords—"cocksucker" and a few others. He was making me scared by my hearing him being scared. I tried to cover my ears so I couldn't hear him, but still [I] wanted to know what was going on. I heard him say [that] we lost the engine on the left wing, so I looked up and out the window and saw what looked like a hole and [I] screamed. The stewardess told me to keep my head down. The pilot and copilot were discussing something to get control of the plane and [they] said something like, "Are you sure?" and "We have to try it." The stewardess then started screaming at the pilot to land the plane anywhere, just to land it now! The pilot came on and told her, "Amy, baby, calm down."

I started yelling, "Just crash the plane now. I can't take it anymore not knowing what is going to happen." All the time I was thinking, "We are going to crash into a hill or a tree." Heidi and I were holding hands. She was trying to calm me down. The pilot mentioned Memphis, and I told Heidi, "I just came from there. We're going in the wrong direction." I heard one of [the cockpit crew] say, "We can't keep circling. We have to land this plane." A voice came on repeating, "Flaps up . . ." something, something. I couldn't understand the rest. We were then told we would be landing at the next airport. I thought if we didn't crash in the air, we were going to crash when we hit the ground or we would flip over when we hit. One of [the cockpit crew] came on and told us we would be crash-landing at full speed [and] to brace ourselves and keep our heads down.

MALE, AGE FIFTY-ONE, SEAT 5A

The aircraft was still climbing through the clouds, and I was reading my book when the plane was shaken by three or four jolts. For a split second I thought it was just severe air turbulence, but then the shaking resumed and the plane banked to the left. In seat 5A I was in a good position to observe the left engine and also watch the pilot and copilot fight for control of the aircraft. The engine sounds began to rise and fall—no longer a steady cruising-speed sound. The pilots recovered from the left bank and momentarily leveled the plane. Then the aircraft banked sharply to the right. In fact, the bank to the right was so steep, I thought the plane was going to roll over. Once again the pilots brought the plane back to level flight for a few moments, and then it went into a violent bank and dive to the left, later described as a "spin." Sometime during these maneuvers I recall looking out and seeing the left engine and prop "shaking" the wing.

Many of the passengers were yelling or screaming by now, and it was very disorienting as we were still in the cloud bank and there was no frame of reference for what was going on. Either shortly before or after the spin started, the left prop began to come apart. I could see the propeller blades breaking loose and flying back over the wing, and then the engine cowling was torn loose and went back over the wing as well. Along with everything else, I was concerned that the debris was going to tear up the control surfaces of the wing and tail assembly. When the plane was finally stabilized, I could see that only one-half of one propeller blade remained on the left wing.

After the plane stabilized, the stewardess yelled at everyone to assume the "crash position" and for the passengers near emergency doors to read the instructions. We were still in the clouds, so I didn't know our altitude, but from the changing air pressure on my ears I knew we were gradually going lower. . . . It seemed like forever, but we finally came out of the clouds and could see that we were still flying over rural farmland. We could only assume that the plane was headed for an airport, but didn't know where. Finally, the pilot said to tighten our seat belts and to maintain crash position, because it was going to be a "rough" landing.

FLIGHT ATTENDANT AIMEE YATES

We pulled out of the spin at approximately five thousand feet. Immediately Ray began to look for an airport. We were yawing hard right. The cabin was a mess, but the cart remained with the brake on, in the back of the plane. We had dropped, spinning, thirteen thousand feet in a matter of seconds. I could hear ATC [Air Traffic Control] on the radio talking to Frank and Ray. They had regained control of the airplane somewhat, but it was not flying very well. We were in the clouds, the weather was rainy, and there was no visibility.

I decided now we were not going to die, just crash and get hurt real bad. I was afraid that it would start to spin again and we would end up upside down and burn. I didn't know how long we had, so I began to shout at the people, "Bend over. Heads down!" so that they would be in the most protective position for impact.

I could hear Ray and Frank talking on the radio. I got on the PA [public address] and gave instructions to the people seated at the exit rows to look at the emergency cards and their respective exits and to know how to open the exits when we landed. I told them firmly and directly, and they began to follow my instructions immediately.

I told everyone else to be aware of the closest exit to them, and where they were located in the aircraft. I didn't have my manual accessible to me. This was all from my memory. The plane was too unsteady to stand and retrieve it from the first overhead bin. I was afraid we would spin again and I would be thrown.

I told them [the passengers] we were looking for an airport to land and to check their seat belts and to make sure they were tight. Everyone tightened their seat belt.

I told them their heads needed to be as far [as possible] into the seat in front of them, because their head was the hardest part of their body, and this would keep them from breaking their noses.

I told the passengers to take off their glasses so that they wouldn't poke out their eyes and so that they could see to get off the plane. We were going to get off!

I told them to make sure that their hands and arms were wrapped around their legs and I showed them the brace position from my jump seat. I told them this would keep them from breaking their fingers, so they could get their seat belts off when we stopped. The passengers were really scared, and I was trying to reinforce that we would be OK. I kept shouting, "Bend over! Heads down!" People kept popping their heads up to open the window shades, to see if they could see anything. The window [shade] in row one was opened and all I could see was clouds.

There was one passenger in the hatch exit row [who] wouldn't get his head down, and he had his hands on the ceiling. I continued to yell at him. I realized maybe he couldn't bend over, so I told the people if they couldn't bend all the way over to grab the seat in front of them, keeping their head in the seat and to make sure their heads were down. I continued to yell, "Bend over! Heads down!" Ray said politely from the cockpit, "Aimee, hon, can you be quiet for a minute, please?"

They were trying to find an open airport. During this time I could hear another pilot's break-in conversation with Frank and ATC. We had an emergency and this asshole started talking over Frank. Ray began yelling, "Get off this fucking channel." The other pilot interrupted twice. ATC had given [us] directions to an airport that we didn't have [guidance] for. I heard Frank and Ray say that they didn't have a [guidance] plate [for that airport]. They were still looking for the closest possible airport. . . . We were losing altitude and the aircraft was wobbly.

I could hear a change in the airplane on the left side, a high-pitched noise. I tried to call the cockpit to let them know something was changing. They didn't answer the interphone. So I yelled into the cockpit that there was a funny sound coming from the left side. Frank said something like, "We know." I wanted to know what was going on and to let them know the left side was making a different sound, even though [the engine] wasn't running. I know [the engine] was off, because it was very close to my jump seat and when it is running it is very loud. Frank said, "Aimee, we are a little busy right now."

I replied, "Frank, land this fucking plane."

He said, "We're trying."

I continued to yell, "Bend over. Heads down!" when people would pop their heads up. Finally Ray said, "Aimee, it's going to be OK, we're going to be OK, we have an airport."

I told the people we had found an airport we were trying to make it to. I told them under no circumstances were they to open any exit until we had completely stopped and I told them to. I also told them to remain seated in the brace position until we stopped.

At this time I told them that the cart was out and the brake was on, but that it probably would not hold. I told them under no circumstance were they to try and stop the cart from rolling down the aisle at me. I told them that the cart was for me to worry about. I told them to make sure . . . their arms, elbows, legs and feet were out of the aisle, so that they wouldn't get broken or taken off by the four-hundred-pound cart.

I began repeating, "Bend over. Heads down!"

I heard Ray say, "Frank, do you see it? I can't see it!"

Frank replied, "No, I don't see it."

Ray said, "Fucking *land* it! Aimee, here we go!"

I shouted, "Bend over. Heads down."

I was in the jump seat the entire time, looking at the cart, afraid it was going to roll forward and smash us, giving commands and trying to assess how and where we may land if we didn't make it to the airport.

I saw outside the window as we approached Pine Bluff airport a pond, then trees and finally a field. This was the first sight of the ground I had since we left Little Rock. We came in still yawing hard right and going very fast. I heard all the audio warnings going off, everything from "Trim Fail," "Pull Up," "Minimum," "Flaps," to a second stall warning and the stick shaker.

FEMALE, AGE FORTY-TWO, FRONT AISLE SEAT

We hit the ground and bounced up and back down again. We kept bouncing a lot, and then we stopped. We were thrown forward and back. The stewardess told us all to evacuate the plane.

I can't remember a lot of what happened, but this was the worst nightmare I have ever experienced, and now I never want to fly on another plane again. The pilot told us later when we were talking that he thought we were all dead and that he killed all twenty-eight passengers on board. He said he had no control over the plane when we were going down. I didn't know until later when he was talking to us that we were in a spin.

MALE, AGE FIFTY-ONE, SEAT 5A

The plane was getting close to the ground when it banked slowly to the right. I was afraid the right wingtip might dig in and cartwheel the plane, but the plane leveled out. Just before touchdown, the stall warning started beeping, but I figured that was the least of our problems.

The plane touched down very smoothly and I thought, "Hey, this isn't so bad." Then we ran off of the runway and into a field. I kept expecting a "crash," but we never hit anything. The plane hit the plowed part of the field and just shuddered to a stop. As passengers we were pretty stunned, because although the plane had stopped, the right engine was still running. Some of the passengers started to clap. I jumped up and spun around to seat 6A. The male passenger there hadn't done anything yet, and I yelled at him to "get the door." Together we jerked the emergency door in and then threw it out on the wing. He had forgotten to take off his seat belt and sort of half fell out the emergency door and rolled down the wing. I was right behind him and could smell jet fuel when I put my hands on the wing and jumped down two or three feet to the field. I took off running and didn't look back for ten or fifteen strides.

FLIGHT ATTENDANT AIMEE YATES

We touched down on the runway and bounced up and hit the runway again. [While we were] still going very fast, the passengers began to clap and pop their heads up. I yelled, "Get your heads down!"

Ray said, "Hold on, hold on, Aimee. Uh-oh. Hold on! Here we go!"

We swerved on the runway and continued off into a field. I could feel the difference when we hit the mud. We were no longer on the runway. Mud was flying on the right side of the plane, making it hard to see out of the window. I am not sure how far we traveled until we went up again, like over a hill or something, and crashed.

There was a strong smell of fuel, and smoke was visible on the right side of the aircraft. . . .

I screamed, "Unfasten your seat belts and get out!"

I heard Ray tell me to open the main cabin door. I was trying to get it open. The people were coming forward so Ray went back into the cockpit. I opened it and when I looked out there were people jumping off the wings.

The stairs take a long time to unfold and people began to jump from the main cabin door. People were evacuating from every exit. I saw people running everywhere away from the plane. I yelled for them to run far away from the plane. Ray was coming out of the cockpit window with the escape rope. Frank was in the door of the airplane asking if everyone was off. There was one lady left on board. She was stuck in her seat. She was elderly and had frozen. Frank brought her out and helped her down the stairs. The passengers had evacuated the plane in five or so seconds.

There were so many people running around I was trying to count to make sure we had everyone and keep them away from the right side of the plane. The wing on the right side was broken. Fuel was gushing out and there was smoke. All the emergency exits had been used and were lying on and around the plane. With the smoke and so much fuel spilling out we were afraid it would blow up. I was trying to make sure nobody was hurt [or] got hit with a flying soda can or some other loose object from the cart.

We were running across the field. Some people, including [me], ran out of their shoes. The field was muddy. We regrouped at the end of the runway where there was some emergency trucks and cars. I'm not really sure what or who they were. The people took us in the back of a pickup truck back to Pine Bluff's regional airport terminal, where Ray and Frank went to call the company to let them know where we were.

Sometime after we got to the terminal someone asked me if I was OK. Apparently, I was limping. I had hurt my ankle. A paramedic came over to look at my foot and wrapped it up. I already had ice on it. Later I went to the bathroom and lifted up my dress and saw blood running down my leg. I looked around and found a bleeding cut on the outside of my leg and a huge welt. I cleaned it up in the bathroom and went to show Frank. He called the paramedic guys over again, and they put a gauze bandage on it and told me I needed a tetanus shot. It seemed I was the only injured person [who] received any medical attention in Pine Bluff.

Everything went so fast that it is hard to remember everything I saw and heard.

We were all very thankful and *very shook up*.

3. WASHINGTON, D.C.

January 13, 1982

Air Florida Flight 90

Air Florida flight 90, a Boeing 737-222, departed Washington's National Airport from runway 36 in a heavy snowstorm at about four o'clock in the afternoon. Heavy ice on the aircraft's wings spoiled lift. Flight 90 struggled to gain altitude. It sank steadily, however, and crashed approximately one half mile from the end of the runway into the barrier wall of the northbound span of the 14th Street Bridge, between the District of Columbia and Arlington County, Virginia. It then plunged into the ice-covered Potomac River, coming to rest on the west side of the bridge. Of seventy-four passengers and a crew of five, four passengers and one flight attendant survived. Four motorists on the bridge were also killed.

FLIGHT ATTENDANT KELLY DUNCAN

We boarded the aircraft and we got everything situated. We took drink orders and stored all the articles and the belongings and everything and then Captain Wheaton came over the PA and said the airport was closed for about twenty more minutes [due to snow and the conditions on the runways] and that we would be sitting there for a while. So they came and deiced the aircraft, and everybody was asking, "What are they doing?" You could hear them spraying [the deicing glycol] over the top [of the aircraft] and everything. And we just sat there, and I was talking to the passengers in the back [who] were sitting back there, and the delay was longer and longer and longer, and they came out and deiced again.

And then we were getting ready to leave, and the senior flight attendant came over the PA and said to prepare for departure. I armed the back [door] slides and got my demo equipment together. We still hadn't pushed back, and then we were sort of moving, but we hadn't actually pushed back, so I went up front and both the flight attendants were in the cockpit, and I said, "What are we doing? Why haven't we moved?" And they said that we needed a tug [vehicle that pushes and pulls aircraft], we're too heavy for this tug. So I went back to the back of the airplane and disarmed the slides, and they got a new tug and put the new tug to push back, and we taxied out to the runway.

COCKPIT VOICE RECORDER

FO: It's twenty-five degrees outside. It's not too cold really.
CAPT: It's not really that cold.
FO: It's not that cold—cold, like ten with the wind blowing, you know.

Several minutes pass as the CAPT starts the engines and the cabin crew goes through the takeoff checklist.

FO: Boy, this is shitty. It's probably the shittiest snow I've seen.
CAPT: Go over to the hangar and get deiced. . . .
FO: Yeah, definitely. It's been awhile since we've been deiced.
FO: That [guy] over there. That guy's about ankle-deep in it.
FO to FA: Hello, Donna.
FA: I love it out here.
FO: It's fun.
FA: I love it. The neat way the tire tracks . . .
FO: See that guy over there, looks like he's up to his knees.
FA: Look at all the tire tracks in the snow.
FA: What does N stand for on all the aircraft, before the [registration] number?
CAPT: U.S. registered.
FO: U.S. United States. See, every one of them [has] an N on it. See? Then you see somebody else like . . . ah . . .

CAPT: C is for Canada. Yeah, I think, or is it Y?
FO: I think, I think it is C. There's, ah, you know. Venezuela. Next time you have a weird one, you can look [it] up. Stand by a second.
CAPT: Tell you what. My windshield will be deiced. Don't know about my wing.

Six minutes pass while the aircraft leaves the deicing pad and joins the line of aircraft waiting for takeoff. The captain and first officer discuss their views on deicing of aircraft.

FO: Boy, I'll bet all the school kids are just peeing in their pants here. It's fun for them, no school tomorrow, ya-hoo!

The crew moves the aircraft ahead in the line of airplanes waiting for takeoff, and they go through a series of checklists.

FO: Cockpit door.
CAPT: Locked.
FO: Takeoff briefing. Air Florida standard. Slushy runway. Do you want me to do anything special for this or just go for it?
CAPT: Unless you got anything special you'd like to do.
FO: Unless just takeoff the nose wheel early like a soft-field takeoff or something. I'll take the nose when off [the ground] and then we'll let it fly off. Be out the three two six, climbing to five [hundred feet], I'll pull it back to about one point five-five, supposed to be about one six, depending on how scared we are. [Sounds of laughter.]
CAPT: Ladies and gentlemen, we have just been cleared on the runway for takeoff. Flight attendants, please be seated.

The crew goes through a final series of checks.

CAPT to FO: Okay, your throttles.
FO: Okay.

MALE, SEAT 21D

We rolled out to the runway, and it seems like we were in line for some time, and I know the guy sitting next to me commented that a lot of flights were coming in and landing at the same time. So apparently we were waiting there in line and once [the pilot] hit it [the throttles]. . . it seemed like it ran and ran and ran [down the runway without taking off]. I've flown quite a few flights, and it's not normal. It seemed like an excessively long time. Then he got somewhat off the ground, but like 737s and 727s, you know, normally when they go, they go! . . . It was a real low climb and I guess [there was] no telling how long [it took]. My mind is not clear on time.

MALE, SEAT 18C

They pulled us out [on the runway], and the pilot may or may not have said anything, like, "We're cleared for takeoff."

We started down the runway, and I . . . had a couple of observations. First off, one of the things I've always liked since I was a baby about flying is that feeling of acceleration which pushes you back in the seat. As we started to roll, I was leaning forward talking to Jane [this passenger was traveling with a colleague named Jane], and the [power] came in so smooth that it never displaced me in my seat. It seemed kinda unusual because those little 737s are scooters. They get movin' and they go up.

I'll come back and talk about my flight experience later, but I will tell you that I have a sense of feel that any pilot develops over a period of time as to what is normal. The first thing that ceased to be normal . . . we started down the runway and we were running a little bit slower than what I normally expected, and I figured well . . . that pilot's just on that ice. He's not giving it all he's got [thrust on takeoff]. And we ran down the runway and we ran down the runway 'til eventually we began to bounce like a pickup truck running through an icy road with big potholes in it. The airplane was bouncin'. And we reached the point where I thought we were close but not quite ready to rotate [take off]. We kept going down the runway and the bouncin' got more and more severe, and I turned to Jane and said, "Jane, we're in trouble. Something's wrong. We're in trouble."

And eventually, I wouldn't say the [pilot] rotated [the aircraft]. I'd say he flew off the runway but he didn't fly up. It was not a normal departure. And as soon as he left the ground, that airplane began to shudder. And I would say that it was oscillating in the area of a couple of hertz—

extremely. And this time I turned to Jane and I said, "We're not going to make it."

Looking through the [window] I could see the treetops over on the left-hand side. [They] were higher than I was. And at that point I had no doubt whatsoever that we were going in.

Just before we ceased to be on the runway—and I say that because we weren't flying in my opinion—I felt momentarily like the [pilot] was coming back on the power, and that was only for an instant, and then I felt more power being added, and whatever [power in the engines the pilot] had left he put it all in there at that point in time, and that was put on about the point where we drifted off the end of the runway. He threw everything he had and there was an additional amount of power above and beyond anything from an aural point of view, and for the first time I felt a slight . . . amount of acceleration. That acceleration maybe [was more] in my mind than it was in fact.

I felt an instantaneous power change, and I was saying to myself, "Thank God. That SOB's gonna give us a bath!" OK? And I expected them to. What I was kinda expecting was that we were gonna start winding down the approach lights, the approach from the other end, and that kinda scared me because I figured that's gonna tear the hell out of us and give us a nice fire, and I was assuming that maybe what this guy was gonna do was . . . I knew it was icing, and [the pilot] didn't have much option. And I know he knew that he was out of runway, that he couldn't do a normal abort [of takeoff] of any kind, but I really seriously was thinking that [the pilot] was somehow gonna get that thing off into the lagoon, which was kind of an open lagoon that sits off to the left side as you come down there, and there was this momentary change in the power setting, and then maybe a half a second later or something like that, then I felt him bringing in all the power he had . . . .Now what that pilot might have been trying to do, I don't know. I was hoping that he was going to abort and take us into the water.

COCKPIT VOICE RECORDER

Sound of the engines spooling up.

CAPT: Holler if you need the [windshield] wipers. It's spooled. Real cold, real cold.

FO: God, look at that thing. That don't seem right, does it? Ah, that's not right.
CAPT: Yes, it is. There's eighty [knots].
FO: Naw, I don't think that's right. Ah, maybe it is. . . .
CAPT: Hundred and twenty [knots].
FO: I don't know.
CAPT. Vee one [takeoff decision speed]. Easy. Vee Two [climb speed].

Sound of stick shaker warning of an impending stall.

CAPT: Forward, forward. Easy. We only want five hundred [feet]. Come on, forward. . . . Forward. Just barely climb[ing].
CAPT: We're [falling].
FO: Larry, we're going down, Larry.
CAPT: I know it.

Sound of impact

FLIGHT ATTENDANT KELLY DUNCAN

I was sitting there thinking, "Get this baby up!" 'cause we were not going up as fast as we should have been.

. . . We did not go into a steep [climb] that we usually do [on takeoff]. We did not have that many passengers on board. We didn't go into a steep [climb]. And [the sound of the engine] was not loud. Usually I sit there at takeoff and put my fingers in my ears. It was not as loud as usual. And we pulled up. A few seconds later [the airplane] started to vibrate even worse and worse and worse, and the man sitting in the front in [seat] 21D turned around and looked at me with a horrified look on his face, and I just looked at him trying not to look scared myself, which I was. And the next thing I knew I was in the water.

MALE, SEAT 21D

Then the airplane started shaking like it was shaking apart, and I happened to glance out the left side of the window and it appeared like we were coming back to the highway.

I remember it started vibrating.

. . . [The impact] was obviously quite violent, as sore as I am right now. But at the time [I didn't realize it]. It's strange, you know. I can't really describe [the impact]. You know I felt the vibration. I felt it hit the water. I didn't know we hit the bridge. I guess I was outside before I really started figuring out just how bad things really were. I guess the first question I kept asking myself is, "Am I really alive?"

I can't really describe it. You know, you knew it was over. I guess it's the best way to describe it. I looked at the guy in front of me . . . and he kind of shook his head, and I reached down and tightened my seat belt one more time as tight as I could get it. Next thing I knew I was in the water.

MALE, SEAT 18C

I told Jane, "No! Look! Just do what I do. We are not going to make it. We're going in," or something like that. "Just do what I do." And I put my arms up around my head, and I put my head down in my lap, and I looked to see if she was doing the same thing. As I said, the nose [of the airplane] went up. It appeared to me that we were a little bit of wing-left-low. . . . It seemed like we were falling off on the left wing, and I stopped worrying about Jane, or anything else.

I just had my head in there [in my lap] and I was waiting for the impact.

It didn't take long, maybe a second after that. The impact when it came surprised me. I mean, it was a significant impact and felt like . . . uh, maybe bending, you know? I don't know how to draw a comparison to anything. . . . In the past I have been rear-ended [in automobile accidents] a couple of times and it was obviously more of an impact than that. But I was still conscious. A second or an instant later I felt another impact, and I felt myself black out, and as I blacked out I just said, "God, help me." And I figured that was the last thing I was ever thinking. I don't know how long after that, I was cold, cold and wet.

As I blacked out . . . I knew exactly what was happening. [I] blacked out, and I figured that was it.

MALE WITNESS

I got off from work at approximately 3:00 P.M. We were headed north on the 14th Street Bridge. As we approached the bridge, about 150 feet across the bridge, coming from Virginia, the only thing that saved us was a disabled car in the right front lane. Traffic was backed up. We got into the flow of traffic and changed lanes to the extreme left.

We felt a strong draft. It felt like someone turned on a fan. Then the plane hit the truck we were riding in. There were three of us in the truck, and I was next to the door on the passenger's side.

We sat there momentarily. Then I said, "Let's get out before [the truck] falls over the bridge."

We all got out of the big Ford truck; [it's] like a dump truck. It has a crane on top and ten wheels. It was braced against the bridge on five wheels.

We didn't see or hear anything out of the ordinary that was about to happen, because it was snowing.

FLIGHT ATTENDANT KELLY DUNCAN

I don't remember taking off my seat belt. I don't remember anything except . . . just swimming to the [surface].

I had the sensation of . . . I felt like I was dying. I felt like I was in white, and I was thinking, "This is what it's like to die" . . . 'cause I knew there was a problem but I thought I was dying or dead and then I was in the water and I swam to the [surface].

. . . It was quiet when I came up to the [surface] of the water. I think I was the only one screaming for help. Everybody was kind of in a daze. The only thing I heard was that other girl tell me she couldn't find her baby—that was the only one.

I swam over there. . . . I am not a strong swimmer. I'm an average swimmer. I don't think even the strongest swimmer could swim in that water. There were glaciers in there. I was so cold . . . .[Under normal conditions] I could have . . . swam from the wreck to the shore 'cause it wasn't that far. But I couldn't swim because I was so numb. I could not function. Fingers were blue. I could feel the blood freezing in my fingers. I looked at my hands and the blood was stiff. I couldn't move my hand on the fuselage because my fingers were sticking to the [metal surface]. . . . I kept sticking my fingers in my mouth to try to keep them warm. I had a real hard time swimming over there. I saw over there, and there was a hole about that big [gestures with hands] in the fuselage, and I grabbed on to that and tried to pull myself out of the water, and I couldn't pull myself out and there was another girl out there. She was quite a ways out in the water, and she was saying, "I can't find my baby, I can't find my baby. Please help me find my baby. I can't swim."

I couldn't let go [of the part of the fuselage she was hanging on]. I said, "Swim over here. You can swim over here."

And she said, "I can't, I can't, I can't move my legs. I can't move my legs."

I said, "I can't come over there and get you. You got to swim over here."

So somebody threw her something and got her over there. There were three others hanging on to the airplane where I was, and this other girl. I looked over there while hanging on, and [the girl next to me] grabbed a life vest that was still in the packet, and she just looked at it in a daze, so I grabbed it from her . . . and pulled it open with my teeth and pulled out the vest and inflated it and handed it back to her, and she just held on to it and then people were coming to get us out, and [the piece of fuselage] was sinking fast. . . . We were all just . . . I was screaming for the people on the bridge to get help, and the girl beside me wasn't saying anything. She was really in shock.

MALE, SEAT 21D

I wasn't aware [that] we hit the bridge until last night when I saw it on the news. Where I was sitting when we hit the water, I don't think I [fully] realized what had . . . happened because I seemed to be conscious. I didn't seem to be hurt. I reached down and unbuckled my seat belt.

It seemed like we were at sea level, or about at water level. It seemed like I was in the water up to my waist. . . . I think—I wouldn't swear to it—I was lying on my side. I think the seat came loose, but I really don't know at that point. All I wanted was out.

. . . I saw an opening out the back—because the back was gone, so I went out the back. [I] found a piece of metal to hang on to and found something underneath [the water] that I could stand on, so I kept trying like to run in place and hang on to the thing, so I started yelling to the people on the bridge to get somebody to help us.

. . . I don't think it really dawned on me that the rest of the people weren't getting out just like I was. I saw three or four people over on my left side. I think [one was] a stewardess. Some lady had kind of been thrown away from the airplane, and one of the guys was finally able to reach her and help her get up close. I didn't know if that was the stewardess. It didn't seem like she was—best I could remember. One kept saying his legs were broken.

. . . I guess I was probably in the water twenty or thirty minutes. I really have no—it was too long however long it was. And finally they got me out and here I am.

. . . In my opinion it was the grace of God that I got out of the airplane, and why I turned back or why I did anything at that point, I can't give you an answer for it.

MALE WITNESS

The airplane was totally demolished. The rear end was on top of the ice. The part where the engine is and the front were completely submerged. After looking for a moment, you couldn't see nothing. But you could hear people screaming. You could see luggage and other debris everywhere. Then three or four minutes later, you could see a man emerge from the plane. You could see his head. I couldn't tell whether he made it or not. He was holding on to the plane and everyone was telling him what to do. Then we could see some more people coming out of the wreckage. I could see the airplane sinking. So I don't know if any of them made it. I know there had to be some survivors, because I could see them holding on. I saw a man . . . he was on the George Washington Parkway side, not involved in the accident, I saw him walk out to the water and begin to crawl along the ice in an attempt to get to the airplane to rescue some people. The ice was probably thick enough to [stand on].

After we got out [of the truck], we walked around and looked around. We could hear people hollering and crying in the airplane, and we heard people on the bridge saying, "Oh, my God!"

We saw a lady in a brown Mustang sitting in the passenger seat.

She seemed to be OK. She was asking for someone to help her husband and saying, "We can't let him die like this!" But then we looked at her husband. The top of his head had been cut off completely. The blood was running very heavily out of the car and out of the door. Some guys pulled her out of the car. Both the deceased and his wife were black. People were crying both on the ground and in the airplane. We saw one car [that] was smashed flat.

There were so many people there. That's about all I can remember at this time. Plus, I'm very upset now, and my back is hurt. All I can say is we're lucky.

FEMALE, SEAT 17F

The only thing I remember was, I was holding on to the plane, and I could hardly hold on. I must have been holding on, but my hands and everything were so cold that I had to hold everything tighter to make sure I was holding on to it. There were people on the shore saying everything was going to be all right, just to hold on, and that help was coming.

And out of nowhere I remember seeing a helicopter, but I never saw it come down or anything for [the stewardess], but I saw . . . I looked up and I saw somebody—I don't know if it was the stewardess or not—I saw a body going up and holding on to the rope from the helicopter, and I thought to myself, "I'll never be able to do that. I'll fall." I don't know if [the person being pulled up inside the helicopter] . . . had [the rope] tied around her leg or her body. I just thought she was holding on to a rope, and I figured I won't ever be able to do that, 'cause my hands, everything, was frozen solid so that I would not be able to hold on to the rope, because I wouldn't be able to feel it, you know, or anything.

Then the helicopter came. It was a dark color. I thought it was green. But now they tell me it was blue. It came down, and I remember it came down really close, and I thought to myself, "Why is it coming so close with the wind and the snow and the water—the water was going everywhere and he was going to make everybody lose their grip and everybody's going to drown."

Then I put my head down, and that's the last thing I remember till I was in the hospital. I don't remember anything after that. . . . I watched it on TV and everything, but I don't remember it at all.

MALE, SEAT 18C

A few minutes later, or some amount of time that passed, I don't think it was much [time], cold water, which was at least up to my waist at that point in time, maybe up to my chest area, I'm not sure—revived me, and at that point, I was in a momentary panic, and I consciously said something to myself like, "Control!"

And then, um . . . I'm kind of walking through the numbers in my mind. I said, "Number one, get the seat belt." So I got the seat belt. "Number two, get Jane's seat belt." And I reached over and I think we must have still had our heads above water at that point in time. I helped her undo her seat belt.

The reason I think I still had my head above water . . . I had no visual impressions from inside that airplane after the crash. I had air in my lungs, and I think any air that was in 'em probably was knocked out of 'em, but I was . . . I had air. I undid Jane, and I was thinking, "Go aft!" because I felt if there was [a way out] it was probably going to be in the rear if the tail [had broken off]. And . . . um, I got hold of Jane, and we kinda struggled. I would say we moved maybe the distance of the equivalent of two rows of seats. And I don't know. . . . My legs hurt. I had to fight to get 'em loose. [Jane] was fighting, and I was kinda, uh . . . She was kinda struggling, and we did get our legs free, and I was pushing her and jointly we kinda worked our way what I felt was aft. I don't know what my orientation was, to be honest with you. And it took us maybe thirty seconds, maybe fifty seconds, something like that, to reach a point. It was dark initially, and we reached a point where it was light, and we got up above the water. And as we surfaced the two of us were there kinda holding on to each other. And I looked around and I couldn't believe what I was seeing.

I . . . um, as I came up, my head was more or less facing north, I think—north or northeast, kinda looking away from the airport in the general direction of the Washington side of the river. I looked around and I saw a few people in the water . . . and I was about as far away from that piece of fuselage that I held on to for most of that period as I am from this gentleman now [the NTSB Survival Factors Group interviewer].

I . . . uh, I said to Jane, "Come on. Let's get over there [to a floating piece of fuselage]." And she said she couldn't swim, and I said, "My legs are broken." And she said, "Mine, too."

And we kinda played this game. You know when you're a kid you play Snap the Whip? So it was [holding hands in a line]. There was lots of ice giving some kind of sustaining [flotation]. I mean, you know, it kept you a little bit up at the surface of the water. And I struggled up enough where I had more or less my upper body above this much [he shows with his hands]. My arms were free, and I kinda pulled [Jane] and then pulled myself by pulling back, and pulled her, and we kinda dog-paddled and snapped our way over to that piece of fuselage. And I pushed her up or tried to push her up on it. And I did a head count [of passengers whom he saw in the water]. There was a stewardess in the water. I didn't know she was a stewardess at that time, but the gal [the stewardess] was in the water south of me. I was at that time, by the time I got there, I was looking south at National Airport. Jane was to my left and I grabbed on to that piece of fuselage. I noticed that the wing—something that surprised me was that the wings were [pointed] south. And we had taken off going in the other direction.

And I heard some noise back upstream—screaming—from where I'd come from. I hadn't noticed [it] originally. And as I climbed up on this piece of [fuselage] there was a man in the water right in front of me, but the fuselage was really ripped up. And just the other side of this piece of fuselage there's a guy and about this much of him is pokin' through the water—head, neck, and tops of his shoulders, I'd say. And I said, "Are you all right?" He said, "I'm not gonna make it." I said, "Yeah, we'll be all right. Somebody'll be here. Don't worry about it." He said, "I'm strapped in, and I can't get loose." I said, "Don't worry about it. We'll be all right."

Someone upstream screamed, "Help! Help! Help! Where's my baby?" and stuff, you know, for a moment. I mean, I looked at her and at first I asked Jane. I said, "Jane, if I hold on to this and hold on to you, you think you can get out there far enough [to] get a hand over to her?" Jane said, "No. My, my arms, I think my arms are broken. My legs are broken." I said, "OK. How about if we do it the other way around? Can you hold on to this if I hold on to you?" And Jane said, "Maybe. I don't know." And then I kinda started to shift myself trying to figure out what was a way to get there [to the stranded woman] and there was kinda like insulation layin' in the holes between the walls of the airplane there at that point. I pulled some of that stuff out. . . . I think that's what it was. And the longest piece I was able to get was about that long—about a foot and a half, maybe two feet. And I kept looking to see if there was anything longer.

I knew I couldn't wait too long. This girl was in hysteria. She was really struggling. And she was fighting against herself more than anything else. So I said to Jane, I said, "Look, just kinda hold on to me so that I can get back to her." So I pushed myself off. I . . . got out to her and she was able to hold on. . . . So I pulled her up to me. She got close. She grabbed my necktie. And that scared me, because she really got a good grab on it. And she said, "My baby, my baby. We gotta go look for my baby." I said, "No. First we gotta get over there." And again I did this thing of working her up toward the fuselage, got her so she was up that way toward the fuselage from me. and then I dropped my legs down and my body came up and I dog-paddled back over there, and she's still holding onto my necktie. And she's still screaming about her baby. And it took me a minute or two to get back over to the fuselage. And she was constantly holding on to my tie, and I kept saying, "Don't hold my tie. Don't hold my tie. Hold my arm. Hold my belt." And I got back over there and I found a ledge underwater. [It] might have been a window ledge or might have been a door or something, but I was able to pin this knee onto something that was part of that fuselage. And by virtue of that [I] was able to raise my body up high enough that I was, my body was kinda leaning against the fuselage. All I had on was a shirt, 'cause I had taken my coat off in the airplane. It was very hot in that airplane while we were still within it. And this gal still had my necktie.

And I eventually . . . kinda got my arm under [hers] and I locked hers around behind me, because I was a little concerned that she was going to pull me off [the fuselage], and if I got away from there, by then it was gettin' so cold I didn't think I could get back. And I held her there and tried to get her to hold on to this thing. She kept wantin' to hold on to me, anyway. Boy, as soon as I knew I had her there . . . I didn't feel too bad. And then I kinda loosened my tie as much as I could. And then I looked at my watch, and it was seven minutes after four at that time.

. . . I looked up at the bridge. I had by now noticed that there were six people [including me, who had survived the crash and were in the water].

. . . This fellow you heard about and one other guy who was about, oh . . . as far away from me as you are again, on the other side of the fuselage and kinda downstream. And he told me later when he came to visit me, he was standin' on a wire and leanin' against the fuselage, but his basic support was a wire he was standin' on.

. . . By now it's seven after four [o'clock in the afternoon] and everybody was yellin' and there was a bus up on the bridge, and we watched everybody get off that bus and come and look [down] at us, and I never knew that there'd been a massive wipeout on the bridge at the same time. . . .After a while they got back on the bus and they left. I said, "I hope they're goin' for help," but I was kinda thinking, "Geez, that [bus] would [serve as] a nice ambulance if they could just get us outta here." But they left.

I wasn't too worried. I figured, well . . . here I am right in the downtown of Washington, D.C. There's not too much to worry about except freezin' to death. And we were yellin' and yellin', yellin', and some people dropped some lines down off the bridge, but they were probably twenty or thirty feet too far south of us to do any of us any good. And I was kinda mentally calculating at this point the possibility of pushing myself off [the piece of fuselage I was floating on]. . . . The stream was movin' that way. Maybe I could float down and snag on to one of those lines. I knew I couldn't find [a line] but I might be able to snag on to one, and that was probably just as good and somebody somehow would get me up [out of the water]. But I had these other people there, and we were yellin', "Hurry, hurry, hurry!" It was cold, and this thing [piece of fuselage] was sinkin'. [A man] jumped in the water [from the bridge] on the other side over near the edge of the bridge and [he] started putting a line out to us. And I said, "Oh, thank God, somebody is smart and brave enough. That guy's bringing a line out here. And if you can just connect that to something we've got a chance."

He struggled and struggled and struggled and about five minutes later I again looked at my watch, and some ambulances showed up, some kind of rescue vehicles showed up on the D.C. side of the water with respect to my body orientation. And I said, "Oh, that's it. Here they are!" But they just sat there and didn't have any way to get to us.

This guy kept comin'. He was struggling like hell and tryin' to go through the ice and water and what have you, and [he] was making very very slow progress. But he was tryin' the whole time, and I mean he was working at it.

And the woman somehow or other got her other arm loose and

got back on my necktie again and throughout the entire period she wanted me to go look for her baby, and I said, "I can't look for your baby. Somebody else is going to have to find your baby for you."

A stewardess and I, we found next this guy who was still in the water. There were some life jackets, little yellow things, Mae West type, sittin' in plastic bags, and they were floatin' in the water beside this guy in front of me, and I asked him to hand some of those up to me. and we found one somewhere else. I don't remember now where we found it. I think it was on our side. We got that open and I got it around the neck of Jane, and the flight stewardess worked her way up and helped her get it inflated. So Jane had a life jacket on, and it was inflated. That made me feel pretty good. The stewardess was just holdin' on and she became quiet. Originally she was doin' a lot of yelling for help and stuff and gettin' the message through [to the people on the shore] that we were cold and scared and hurt and all that kind of stuff. I don't think anybody had any doubt about that anyway, but she kinda held on to her position there. She didn't move too much. I looked over there [to her] every now and then and her eyes were wide open and it was obvious that she was not in real trouble except cold and freezing, as far as I could tell.

FLIGHT ATTENDANT KELLY DUNCAN

As soon as the helicopter brought me over to the shore, there were men there, and they took me off the rope, and I just let them have me. I just thought, "I'm safe now. They'll take care of me." I was semiconscious in the ambulance, 'cause I remember speaking to the paramedic. They were asking me what my name was.

MALE, SEAT 18C

Jane was in pretty bad shape. Her head was very close down to the water and every now and then she was kinda at arm's length and I would reach over and pull her head back enough just to make sure, 'cause her face was mainly away from me. She had a head gash and she'd been bleeding in her head. I just wanted to make sure her nose and mouth were above water.

. . . I talked to this other fellow [in the water] who was in front of me. I kept saying, "There's a life jacket beside ya. If you can figure out how to get those out [of the plastic containers] and get one on her [the woman who had lost her baby], I'll help you get one on and you can get one on me, and I can probably throw one to that guy over there." And I think there were only three of [the life jackets]. It doesn't matter. And the guy said, "I can't. I can't move. I'm strapped in." And I tried to reach to see if I could reach down this way underneath [the water] and find if there was a hole so I could reach through to him, but there was no hole there. There was no way. I couldn't get across over the top [to him]. I had my body as high [out of the water] as I could get it, and I was out of the water up to about this height, because of where my knee was ledged, and I just couldn't reach either the life rafts or the guy. And he said, "I'm . . . I'm not gonna make it." And I said, "Just hold on. They'll be here. They'll get us out of here. Don't worry." And I kinda kept saying that to everybody else, and then the stewardess was saying [the same things] to Jane from time to time too.

Then about twenty-five minutes after, this helicopter showed up. And when he first showed up, I heard him comin', and my prayers were answered. And I looked around and eventually he came over and [the helicopter rotors] threw a big squall of ice and water. Everything was blowing in our eyes, and we were already cold, obviously. . . . Eventually he circled around this way, and now I'm lookin' south, holdin' on. Here's the [Air Florida] airplane and I don't have a grip with my hands anymore but I'm able to lean and stuff [to hold on], and this guy [in the helicopter] drops a rope with a ring on it. And this guy who was strapped in [to his seat], I knew damn good and well they couldn't get him out without comin' into the water for him. And he knew it. And I knew that probably the person in the worst shape and also a little bit of a threat to me personally was [the woman who had lost her baby] who was next to me. So when [the helicopter] dropped [the life ring] I grabbed hold of it and I tried to get it to [the woman who had lost her baby]. I tried to figure out some way to get her loose [from the fuselage and the water] without lettin' go too. I couldn't. A helicopter man pulled [the ring] back up and [the helicopter] looped back around again. He was yellin' at me, and I don't know what he was sayin'. I couldn't hear him. I could see his face. His mouth was opening, and I was going, "What? What?" And he tried to [explain what he wanted] again.

And the guy in front of me [who was strapped in his seat], I don't know if we talked or not. But the thing we were tryin' to do was get [the woman who had lost her baby] out of there first. Eventually the helicopter guy . . . went and [rescued] somebody else—I think it was the stewardess. And I noticed how much time it took. He might have gotten her before that. Right now [the] sequence in my mind is not all that clear. But I remember watching . . . [and] how much time it took, and I was thinking, "Hell, now I know he's here I got all the time in the world."

4. *CINCINNATI, OHIO*

June 2, 1983

Air Canada Flight 797

Air Canada flight 797, while flying from Dallas, Texas, to Toronto, Canada, experienced an in-flight fire in the aft lavatory. Approximately one hour and forty-four minutes after takeoff an emergency was declared after the flight crew reported smoke in the airplane. An emergency descent and landing was made at the Greater Cincinnati Airport about eleven minutes after the cockpit crew declared the emergency. When the aircraft, a DC-9, came to a stop on runway 27L, airport firefighters began applying a flame-retardant foam to the airplane almost immediately. An intense fire spread quickly and engulfed the airplane's cabin interior. There were forty-six people on board—forty-one passengers and five crew members. Eighteen passengers and five crew members managed to evacuate the airplane in heavy smoke and in spite of the intense heat of the fire. There were twenty-three fatalities. The airplane was destroyed by fire.

COCKPIT VOICE RECORDER

FO: How's your seafood?
CAPT: It's good. Is your steak nice?
FO: What was that? It's right here. I see it.
CAPT: Yeah. DC bus [an electrical circuit indicator in cockpit].
FO: Which one is that?
CAPT: DC bus, the left toilet, left toilet flushing. Better try [the circuit breaker] again. Push 'em in.

FO: Push it one more time, I guess.

Sound of electrical arcing.

FO: What!

Sound of electrical arcing.

CAPT: That's it. Won't take it.
FO: No.
CAPT: See anything else? [There is] nothing on the panel. Ha. Like a machine gun.
FO: Yeah. Zap, zap, zap.
CAPT: Put [it] in the book.
FO: Log it.
CAPT: Now, I want to log it, eh. Somebody must have pushed a rag down the old toilet or something, eh? Jammed it, and it overheated.
FO: Is it flushing? You pushed?
CAPT: It's flushing. Yeah. Toilet flushing, three [circuit] breakers banged.
CENTER: Air Canada seven ninety seven, contact Indianapolis on one three three point zero five. So long.
CAPT: Indianapolis Center, this is Air Canada seven nine seven maintaining three three zero, direct Louisville on course.
CENTER: Air Canada seven ninety seven, Indianapolis Center. Roger.
CAPT: Don't see the ground too often today, eh?
FO: No, a lot of cloud, eh. The whole area. . . .
CAPT: Better have dinner here.

Sound of chime.

FA1: Yes?
CAPT: Sergio, could I try for [my dinner] now, please.
FA1: Sure.
CAPT: Thank you very much. [To FO] Do you want any of that fruit or should we give it to the girls? . . .
FO: No.

CAPT: I don't want it. There you go.
FO: Thanks.
FA2: Excuse me, there's a fire in the washroom at the back. They're just off, went back to go to put it out.

FLIGHT ATTENDANT JUDITH DAVIDSON, AGE THIRTY, AFT JUMP SEAT

[The flight] was uneventful until I walked to the back of the cabin toward the end of the meal service. I was attending to an ill passenger who had been moved to the last row of seats on the right when I smelled something strange. I checked the storage compartment behind row 21, and as I reached across the aisle for a CO_2 extinguisher located on the forward wall of the aft lavatory, the smell was stronger. I opened the lavatory door a couple of inches and saw light gray smoke from floor to ceiling. I breathed some smoke, which made me feel dizzy. [Flight Attendant] Laura [Kayama] then notified Sergio [Benetti] and the cockpit crew. Sergio discharged the CO_2 extinguisher into the aft lavatory while I reseated passengers forward. The first officer came back into the cabin and I notified the captain that passengers were being moved forward. I began to turn on overhead air toggles and reassure passengers.

COCKPIT VOICE RECORDER

FO: Want me to go there [to the washroom]?
CAPT: Yeah, go.
FO: Leave my dinner in the thing there.
FA3: Okay.
FO: Got the breakers pulled?
CAPT: Pardon me?
FO: You got all the breakers pulled out?
CAPT: The breakers are all pulled, yeah.

FIRST OFFICER CLAUDE OUIMET, AGE THIRTY-FOUR

Flight attendant advises [me] of a fire in the back washroom. I go back. [Flight Attendant] Sergio Benetti tells me that he has discharged some of a CO_2 to try to get to the washroom but couldn't because of smoke. I tell M. Benetti to get the passengers all up front and seated, not to use H_2O extinguisher, and to get me the other CO_2 bottle. Meanwhile I will report to the captain and get smoke goggles. I notice that [Flight Attendant] J. Davidson is opening the air vents [above passengers' heads].

COCKPIT VOICE RECORDER

The FO is now back in the cockpit.

FA3: Captain, is it okay to move everybody up as far forward as possible?
FO: Okay, I, eh, you don't have to do it now. I can't go back now. [The smoke] is too heavy. I think we'd better go down. I got all the passengers seated up front. You don't have to worry. I think [the smoke] is going to be easing up.
FA1: Okay, it's starting to clear now.
CAPT: Well, I want . . . hold on then.
FA1: I just can't go back. It's too . . .
FO: I will go back if that appears better, okay?
CAPT: Yeah, that's okay. That's okay, yeah.
FO: So . . .

FO now prepares to go back to the lavatory one more time.

CAPT: Take the smoke mask.
FO: You have control?
CAPT: Take the goggles.
FO: I'll leave the mask on.
CAPT: Okay. Okay, go back whenever you can but don't get yourself incapacitated.
FO: No problem, no problem.
CAPT: Okay.

FIRST OFFICER CLAUDE OUIMET, AGE THIRTY-FOUR

He goes back from the cockpit to the bathroom.

I walk directly to the washroom, and before I open the door I decide to touch it, and it is hot all over. I pause a few seconds, and I decide not to open [it], and [I] tell M. Benetti accordingly. I rush back to the cockpit.

I inform Captain Cameron that the situation is very serious, and we have to go down. He tells me that we have an [electrical] failure, and I notice some lights [were lit up] on the Cautions and Warnings panel [in the cockpit]. I take my seat [in the cockpit] and declare an emergency twice, interrupting an air-ground communication, without success. I squawk 7700 on the transponder, and the third call consisted of a Mayday and that we had an emergency. We had a fire on board, and that we were going down.

COCKPIT VOICE RECORDER

CAPT: Memphis Center, this is Air Canada seven nine seven.

INDIANAPOLIS CENTER: Canada seven ninety-seven, Indianapolis Center, go ahead.

CAPT: Yeah, we've got an electrical problem here. We may be off communication shortly. Ah, stand by. . . .

FA1: [The fire is] getting much better. I was able to discharge half of the CO_2 inside the washroom even though I could not see the source but it's definitely inside the lavatory. Yeah, it's from the toilet. It's from the toilet. The CO_2, it was almost half a bottle and it now almost [is] cleared.

CAPT: Okay, thank you.

FA1: Okay, good luck.

FO: Okay, you got it.

CAPT: Yeah. I don't like what's happening.

FO: I think we'd better go down, okay?

CAPT: Okay.

FO: Okay, I'll be back there in a minute.

CVR goes off.

In the cabin, among the passengers, the first indication of trouble is the appearance of the first officer running back and forth between the cockpit and the lavatory and the flight attendants requesting that some of the passengers in the aft section of the aircraft move to forward seats to get away from the more intense smoke coming from the lavatory.

MALE, AGE THIRTY-SEVEN, SEAT 8E

We were just finishing dinner when my attention was drawn to the problem by passengers moving from the rear half into the front half of the aircraft. . . .

At first sighting, the smoke appeared to be white in color and coming from the rear-left-side toilet. [The smoke] came forward to about the midpoint between the toilet and the over-wing exits. The steward was holding an extinguisher. Also the first officer came back into the cabin to inspect the problem. He shortly returned to the cockpit. The cabin staff asked us to fully open the overhead air vents. Once [that was] completed, the smoke seemed to recede for a short time but within less than a minute it started to increase. As it did so it became thicker and darker. By this time all the passengers had moved into the front half of the aircraft and the cabin staff had removed the dinner trays.

MALE, AGE SIXTY-THREE, SEAT 3E

My first awareness of a problem was the cabin crew's action to move passengers into empty seats around me from the back of the cabin. I turned and saw some white-gray smoke in the aft part of the cabin. I believe the stewardess mentioned that there appeared to be a fire in the waste disposal unit in the washroom—assumed to have been caused by a discarded cigarette.

The first officer hurried down the aisle with a fire extinguisher and shortly thereafter he returned, moving quickly to the cockpit. On looking back, it was obvious the smoke was building and billowing up the aisle, rising from low down to the ceiling. My first smell of smoke is difficult for me to describe, except to say it was a different odor than was apparent at a later stage when the color also changed from white to yellow-brown. We were advised to put our heads down. I had a paper napkin left over from dinner that I held to my nose.

FEMALE, AGE THIRTY-EIGHT, SEAT 11C

All of a sudden the plane dropped. Coffee cups dropped but the coffee didn't. The lights went out and air stopped flowing out of the vents. Then the airplane began a steep descent. As soon as the airplane made the deep drop the smoke came out of the back much faster and filled up the plane faster. The smell was awful. It made me gag and cough. I used the headrest cover to breathe through. During the descent one of the flight attendants was telling passengers where the window exits were and that they weighed approximately seventy-five pounds. All the while the airplane came down extremely fast. During the descent I was bent over toward the middle seat and was breathing through a napkin. I heard and saw the flight attendant talking about emergency instructions but near the end I couldn't see her. The flight attendant was choking while giving the instructions. She referred to the operation of the window exit and made reference to the fact that you had to pull the handle to open the exit. I felt it would be impossible to have read the briefing card at that time.

FLIGHT ATTENDANT JUDITH DAVIDSON, AGE THIRTY, AFT JUMP SEAT

I felt the aircraft descend rapidly. There was a short meeting in the galley with the other two flight attendants. I wetted after-meal service towels in the galley and told Laura to hand them out. Approximately twenty towels were handed out, when the first officer yelled, "Sit down!"

The smoke was very heavy in the cabin.

I instructed Laura to brief able-bodied passengers at the window exits, and I briefed the passengers in 2A and 2B for crowd control. I then sat down in seat 3C. As the airplane continued its descent I left my seat to calm passengers and assure [myself] that seat belts were fastened. I kept my head down as much as possible. I could not get to row 12 because the smoke was too thick. . . . I kept telling passengers to breathe through their clothes [as filters].

Laura left her seat in the over-wing area and sat in seat 3C. I sat on the floor between the forward bulkhead and seats 2A and 2B. I could only see the passengers' knees due to the dense smoke. I thought it would be better to be in a seat. I got up and, using the seat backs as guides, I found my way back to seat 7B or 9B. I could barely see the gentleman next to me. My eyes teared during landing.

MALE, AGE THIRTY-SEVEN, SEAT 8E

It was at this time that the captain commenced the emergency descent. He appeared to reduce all power and deploy full spoilers [air brakes]. The aircraft's attitude was significantly but not violently nose-down. We passed through a cloud layer at what I estimated to be between five thousand and eight thousand feet.

During the descent the smoke became extremely unpleasant. One had extreme difficulty in breathing, despite the use as a filter of two paper napkins left over from the dinner. Surprisingly quickly visibility dropped [in the cabin] to zero. The fumes were acrid and smelt of burning plastic. They made [me] cough extensively. I removed my seat belt and got my head as low as possible below the seat level. I occasionally raised my head to look out of the window. The overhead oxygen system was not deployed.

FIRST OFFICER CLAUDE OUIMET, AGE THIRTY-FOUR

We have commenced [our] descent, and we now have a [Air Traffic Control] clearance [to descend to] five thousand feet. Captain is flying, and I am now strapping myself in. I put my O_2 mask on, and realize that I left my smoke goggles back there [near the lavatory]. I ask [Flight Attendant] M. Benetti to get [them] if he can.

. . . I notice more flags and warnings. [The captain] also tells me that he cannot fly headings anymore because of instrument failures. I advise ATC of Cincinnati Airport at twenty-seven miles and twenty-five-hundred foot ceiling at the beginning of the descent. I have requested to have fire trucks out. Most of the descent is in the clouds to twenty-five hundred feet.

We receive clearance for three thousand feet, still in clouds, then for two thousand feet, and we become visual. I am manually controlling pressurization [to the cabin] and I [turn] two air conditioning packs off. The airplane is depressurizing since I am now opening and closing my side window to clear smoke. I see buildings in haze and a tower with flashing lights at twelve o'clock. We start a left turn at the same time as ATC advises us to make one. Still [I] cannot see the airport. I ask ATC to confirm that the airport is at one o'clock position, and his reply is negative. The captain asks for five degrees flaps, and before I select, I indicate to the captain that the speed brakes are extended. I have not seen the airport at this time. I am now looking for highways because of smoke condition deterioration. We are still flying at 200 knots. We now see the airport at eleven o'clock position. Gears are down. Speed with final flaps is 140 knots.

MALE, AGE THIRTY-SEVEN, SEAT 8E

Just prior to touchdown I sat back in my seat and did up the [seat belt] strap. I noticed that we had no engine noise at all at that point. The landing was very smooth and the braking severe. Although at no time did I get thrown forward into the seat in front of me I was not aware of any reverse thrust being used.

MALE, AGE SIXTY-THREE, SEAT 3E

As we came in sight of the ground we were over a green wooded area. I remember seeing what appeared to be a major expressway and I thought it would provide a place to land if we had to get down before we reached an airport. I remember the airplane making a couple of course adjustments as we approached the airport. I did not see the airport before we landed as the smoke was getting very bad. I heard passengers emitting little cries. There was no panic, and sound in the aircraft subsided as we were on our final approach.

FIRST OFFICER CLAUDE OUIMET, AGE THIRTY-FOUR

Touchdown is normal, and the captain brakes heavily. Meanwhile I select manual spoilers and high reverse. The airplane stops by the [waiting] fire trucks.

Smoke is intense, and I see Captain Cameron's hands pulling both fire [extinguisher] handles, but I do it myself and I make sure that fire agents are discharged.

FEMALE, AGE THIRTY-EIGHT, SEAT 11C

The landing was not bad and not extremely bumpy. The airplane stopped quickly. I heard the window exits being opened. I had some difficulty unfastening my lap belt. I thought it was like a car lap belt with the push-button release. I said, "I'm still here." My companion went by me unaware I was still seated.

MALE WITNESS ON THE GROUND, AGE TWENTY-FIVE

The plane was coming in for what appeared to be a normal landing but upon touchdown the tires blew because of stomping on the brakes. As soon as the plane came to a stop the fire department was at the aircraft. Within fifteen seconds the two emergency doors over the left wing and the main door [were] open. The slide came out the main door and passengers started getting out. At least two people got out on the left wing. When the doors first opened I saw a little bit of smoke. As the passengers were getting out more and more smoke was coming out.

FLIGHT ATTENDANT JUDITH DAVIDSON, AGE THIRTY, AFT JUMP SEAT

When the airplane was on the ground, I went forward. No one touched me. I felt my way to the exit, opened the door and pulled the slide inflation handle. I do not believe that a passenger would have been able to locate and open that exit. I had a feeling of suffocation and I knew there were toxic fumes in the smoke because of the burning sensation in my throat. I yelled, "Come this way!"

I waited three or four seconds, then exited out of the exit. I waited for a couple of seconds at the bottom of the slide.

MALE, AGE THIRTY-SEVEN, SEAT 8E

Before the aircraft came to a stop, I undid my strap and moved rapidly into the aisle and toward the forward exit. It was extremely dark. I was unable to see anything at that time. The smoke in the aisle was denser, seemed more acrid and was hotter than that around my window seat, although at no time whilst I was in the aircraft did I see flames. I was not aware of passing my colleague in his 8C seat and therefore presumed that he had moved aft, since both he and my other colleague had been asked by the stewardess at the start of the descent to prevent passengers from crowding the over-wing exits after landing, whilst the passengers seated in these exit seats were removing the exit panels.

Within about ten seconds of waiting near the forward cabin bulkhead I saw a patch of light coming from the door, and I ran toward it. I then slid down the slide to the ground. Throughout the emergency descent no one spoke except for a word of encouragement to "positive thinking" from, I believe, the stewardess at an early stage and an insistent but calm voice of a male calling to have the door opened just prior to this being done.

Through the descent, there was a lack of panic. I do not recall touching or being touched by anyone.

MALE, AGE SIXTY-THREE, SEAT 3E

When the aircraft came to a full stop, I heard shouting and some commotion at the front of the aircraft but could see very little. I hesitated a moment, being concerned about possible pileup. But I realized there was no sound in the aircraft for some reason and I can't explain it. I raised my hand and touched the underside of the overhead compartment and found it quite hot to the touch. I knew I had to get out fast. I undid my seat belt and felt to see if the passenger in 3D had gone. He was not there. I went to the floor and crawled out on my hands and knees, past one prone form [who] was in the hang-up baggage compartment. I hesitated to consider helping him but he was obviously out. . . . I continued toward the door. I didn't sense anyone behind me except the one in the vestibule. At first it was clear to me that the door was open. Then smoke swirled and I saw the chute and I plunged down feetfirst. I did not see any flames before leaving the airplane. But I had been conscious of heat and smoke rushing to the front of the aircraft. . . .

FIRST OFFICER CLAUDE OUIMET, AGE THIRTY-FOUR

I removed my O_2 mask, and I see a sliding chute coming down on the forward [right] door. I try to leave through the cockpit window but fall back in my seat because of a wall of smoke. I go out through my sliding window, look at the chute, but nobody is coming down, run to the other side to see the captain in his window looking disoriented. I shout to him to come down without results. No more passengers are coming down at this time. I get a fireman's attention to shoot foam at [the captain], which seems to accelerate his evacuation process. We then directed passengers away from the airplane and discovered that some were missing.

FLIGHT ATTENDANT JUDITH DAVIDSON, AGE THIRTY, AFT JUMP SEAT

When nobody came I ran around to the exit [on the other side of the aircraft]. I saw [the first officer] at that time. I also saw three people come down the slide at [that] exit. I saw Captain Cameron as he exited the cockpit window. I helped move passengers away from the airplane. I noticed a three-foot area over the lavatory that appeared very charred. I saw firemen on the left wing. A few seconds later flames went through the cabin.

MALE, AGE THIRTY-SEVEN, SEAT 8E

I recall noting rescue vehicles in place when I reached the door, and when I reached the bottom of the slide I looked up and noted that the captain was halfway out of the port-side cockpit window with the foam being sprayed on him.

MALE, AGE SIXTY-THREE, SEAT 3E

The firefighters were already pouring foam on the aircraft, when I looked back. But when no one else was seen to leave we came to the awful realization that others would not be able to get out. It was about that time that I observed flames through the emergency exits opposite the wing for the first time. A light wind prevailed, tending to hold the smoke in the main entrance and blowing out the starboard door and emergency exit.

One emergency vehicle with first-aid equipment was already near us and the attendant was looking after a man with a cut hand. My face and hands were quite black and I saw a man with soot on his face. An emergency ambulance arrived and the crew invited me on board. They checked my pulse and breathing and administered oxygen right away. They then put me on the gurney. They had also picked up a young lady passenger, and they left with the two of us for Booth Hospital. The handling of this incident by all concerned at the airport and at the hospital was most impressive.

MALE, AGE TWENTY-FIVE, WITNESS ON THE GROUND,

Two firemen had gotten on the left wing and went to the emergency exits. Whether they went in [the airplane] or looked in is unclear. That is when I saw the first bit of fire, which looked like it came from about the right emergency exit over the wing, but it was contained, so it probably was already in the rear.

Then I saw more and more fire. The fire department sprayed on foam and it went out, started again, [and was] put out, started again, put out. It could have been continuous but just inside [the airplane].

The passengers ran about one hundred feet away. Some of them were bent over. By this time I got called to the phone. This is the first time I have seen something like this, and I hope it's the last.

5. *MIAMI, FLORIDA*

May 5, 1983

Eastern Airlines Flight 855

Eastern Airlines flight 855, a Lockheed L-1011, a three-engine, widebodied aircraft, departed Miami at 8:56 in the morning bound for Nassau, the Bahamas. A quarter of an hour later, while descending from twenty-three thousand feet, flight 855 lost engine oil pressure on its number-two engine, which the crew promptly shut down. The cockpit requested clearance from Nassau Approach Control to return to Miami, and shortly after clearance was given and the flight was making a left turn to a heading of 270 degrees, while cruising at twenty thousand feet, the number-three engine lost power. Approximately four minutes later, the number-one engine lost power. Flight 855 was now gliding for six minutes before the crew was able to restart engine number two. The crew successfully landed 855 on runway 27 Left at Miami International Airport. There were no injuries to the 168 passengers and ten crew members. There was no emergency evacuation upon landing.

TRANSMISSION, FLIGHT 855 AND MIAMI AIR TRAFFIC CONTROL

855: Miami Center, Eastern 855

ATC: Eastern 855, Miami Center Radar Contact, continue present heading. You want higher altitude?

855: Affirmative. Like to go to, say, nineteen [thousand feet] or twenty-one.

ATC: Eastern 855, climb to flight level two zero zero for now.
855: Roger.
ATC: Eastern 855 heavy, have you back on radar contact here about 270 degrees heading radar vectors to Biscayne.
855: OK, 270, heading back to Biscayne. Roger.
ATC: Any problem we should know about?
855: We have precautionary shutdown on number-two engine.
ATC: Do you need any special handling?

FLIGHT ATTENDANT RAPHELIA SPISSO MEDLIN

We had already completed the beverage service and picked up the trash in both first and coach classes when the captain's voice came over the PA saying that we had a mechanical problem and that it would be necessary to return to Miami, and that we would be on the ground in approximately thirty minutes. I was standing at the service area when [another of the flight attendants] came up and opened the cockpit door to inquire as to the nature of our "mechanical." The second officer motioned her to get out of the cockpit. This was an unusual response, which fed our curiosity, and by this time the other flight attendants began to gather in the service area. You could feel the tension mounting at this point, and I jokingly said, "Biscayne Bay is a bit chilly this hour of the morning." We all laughed, and it briefly broke the tension. Very soon after, there was a loud bang . . . which sounded like a compression [engine] stall.

FLIGHT ATTENDANT LINDA PALMER SCOPETTA

The flight had been going smoothly. We had been in flight about thirty minutes, and we were in the process of cleaning the cabin, picking up cups, glasses, in preparation for landing in Nassau. The captain then announced over the PA system that at this time we were experiencing some mechanical difficulties, and we were turning back to Miami for servicing.

TRANSMISSION, FLIGHT 855 AND MIAMI AIR TRAFFIC CONTROL

855: Miami Center, Eastern 855, we have some rather serious indications. We have indications that all three oil pressures on all three engines are down to zero. We believe it to be faulty indication. Cannot be that all three engines have zero oil pressure. The quantity is almost zero. However, these are [the] indications in the cockpit at the present time.

ATC: OK, fine. Why don't you turn about fifteen degrees direct to Miami. Maintain flight level two zero zero, or whatever altitude you want to maintain.

ATC: Eastern 855, proceed direct to Miami Airport.

855: Miami Airport, roger.

855: Eastern 855, we just lost our number-two engine.

ATC: OK, number-two. You still have two turning?

855: Negative. We only have one now. We are going to restart number-two engine.

ATC: Your position now seventy miles southeast of Miami, about fourteen minutes out.

FLIGHT ATTENDANT RAPHELIA SPISSO MEDLIN

The next thing, the captain came over the PA. "Flight attendants, prepare for emergency—flight attendants forward!"

The cockpit door opened, and the second officer said, "Prepare the cabin for a ditching. We're going in."

I remember we all looked very hard at each other. Linda [one of the other flight attendants] grabbed an ammonia inhalant and offered me one. The smell made me nauseous. Shirley [another flight attendant] said, "OK, we've all been trained. Just do it! Get your [life] vests!"

I went to my jump seat and grabbed my [life] preserver and tore it out of its container. I remember [hearing] Shirley's voice telling people that their preservers were under their seats, and a lot of people were crouching over reaching for them. I went to the front of the [airplane]. People were screaming, and there was a lot of panic.

Then Shirley said over the PA, "This is just a precaution. Watch your flight attendants." She read through the instructions over the PA—"Arms through loops, pull over your head," etc. This is when ... the passengers [and I] put [our] preservers on. The top [of the preservers] is very small, and most got tangled, because they were fastened in their seats. There were still some hysterical people, but overall they paid attention.

FLIGHT ATTENDANT RAPHELIA SPISSO MEDLIN

I helped a woman with a child in row 11AB and a few other people in the center who were all tangled up in their preservers. I visually scanned [the back of the cabin], and everyone had a life preserver. I then walked back to the [rear of the airplane] and visually checked preservers in the first few rows. I then grabbed five men to brief [instructed in the use of the emergency slide at the doors].

... Next I heard Shirley's voice telling everyone to take off their shoes and to stow loose articles. We did not have time to collect them. The next thing we heard was the captain's voice.

"Ditching is imminent. Flight attendants, bracing position. Prepare for impact."

TRANSMISSION, FLIGHT 855 AND MIAMI AIR TRAFFIC CONTROL

855: We lost our third engine right now.
ATC: OK, you have the other one started?
855: Not yet.
ATC: You have any of them turning?
855: Not yet.
ATC: Just advised Coast Guard, coming toward you. You are twenty miles west of Bimini right now.
ATC: You are forty-nine miles southeast of Miami right now, twelve minutes, over.
855: We don't believe we can make it.
ATC: We got all the help we can coming out as fast as we can.

FEMALE, AGE THIRTY-SEVEN

There was a stunned reaction [to the captain's announcement of an "imminent ditching"]. A few, very few, people began to cry, and one woman was hyperventilating. I only saw two people in my section [of the airplane] who panicked, but one woman was upset as soon as she got on the plane. Most people were unusually controlled. I think most of us were aware that we were responsible for ourselves and to panic would just make the situation worse.

In my section the flight attendants were very little help. Before the emergency I saw three of them attending to the woman who was upset and nervous about flying. They brought her two drinks and a blanket, because she was shaking. But after the emergency was announced, I don't even recall seeing them after they asked for a "responsible adult" to help with the life rafts. I didn't see them again until we landed.

We helped each other. The flight attendant in the rear section apparently stood on a seat and read the instructions off of the [life] vest and demonstrated how to put it on. However, no attendant in my section reinforced instructions or checked to see if people even had vests, let alone if they were put on correctly. One woman couldn't find her life vest under her seat and received no assistance, so she had to crawl up the aisle reaching under other seats looking for a vest.

FEMALE, AGE FORTY-SIX

When we were notified that "ditching is imminent," the stewardess said there would be two life rafts available when we got off the plane. I did not know where the life rafts were or how they were to be released.

Even though the passengers were upset, they did not get hysterical. They turned to each other for comfort and aid when they saw they could not get much from the stewardesses. They listened quietly to all that the pilot had to say. The stewardess in our area did not do much either to instruct or to assist passengers or to dispel their fears. The pilot was the only one doing any instructing. . . . Another passenger [and I] were assisting the passengers around us with life jackets, instructions, and morale boosting. When the able-bodied men were called forward to assist at the emergency exits, I was appalled to note that they were not seated and strapped in for the "ditching." They were all clustered—standing stooped—at the emergency exit to the front left of us. It ran through my mind, how were they going to help us if they were maimed?

FEMALE, AGE FIFTY-FOUR

[I was] very frightened. Everyone began to pray. Some were crying. But as a whole, I think all passengers conducted themselves in a calm manner, except one lady in the very back, and one small child.

I knew the plane was turning, heard noise. The pilot said we are now changing course. We are headed back to Miami. Next, flight attendant said we might have to ditch plane. She seemed to be calm. But as she announced ditching was imminent her voice was very shaky. Pilot's voice shaky as he told all attendants go to front cabin. They all ran to front. As attendants sat at doors two or three of them were crying. One black attendant was very calm, said she had ditched before and knew what to do.

FEMALE, AGE THIRTY-SIX

A steward stood on the seat in front of the movie screen and slowly put his vest on so we could see how to do it. The written instructions on the vest itself were of greatest value to me but only because there was quite a lot of time during the emergency. The initial instruction [before takeoff] was thorough, but of course the passengers, knowing that we'd *never* need those vests, only listened with one ear!

One woman in the back of the plane started to scream. Otherwise the majority of the passengers remained fairly calm. A few passengers got up and changed their seats to be closer to the exit.

Our steward, although visibly frightened—we *all* were—was in command of himself and therefore [he] gave us courage. He gave us instructions in a loud, clear voice. He got several passengers to assist in opening the door and releasing the [life] raft. He helped those passengers who had difficulty with their vests and he reassured us constantly. My one complaint: We were not told what to do when the plane actually ditched—just how to prepare for the ditching. I asked him what was next, but the pilot came on and said, "Ditching is imminent," and our steward had to sit down and buckle up. Luckily, we never needed the rest of the instructions!

I helped the people seated near me who had difficulty with their life vests, but I was afraid to unbuckle my seat belt. The steward helped us all stay calm, and my seatmate kept me from panicking. I had never met her before but we will be friends forever!

FEMALE, UNKNOWN AGE AND SEAT

I am aware of the criticality of the situation and wouldn't have expected anything other than the necessary cryptic information from the cockpit, but from the cabin crew, some data, such as "the plane will glide"—some of us with unmechanical minds thought [the airplane] might have dropped from above the clouds directly into the ocean when the pilot said the ditching was "imminent"—"the plane will not sink immediately," "we will get the doors open," "there are boats out there ready to rescue us"—[these announcements] would have eased the indescribably thirty-plus minutes of terror, somewhat.

TRANSMISSION, FLIGHT 855 AND MIAMI AIR TRAFFIC CONTROL

ATC: Seventy-nine hundred feet, descending slowly.
855: We have an engine going now. We believe we can make the airport now.
855: We believe we got it made.
ATC: Fantastic!

FLIGHT ATTENDANT RAPHELIA SPISSO MEDLIN

I reached up and took the plastic cover off the emergency handle [on the door] and stowed it in the seat-back pocket [of] 18A. I sat down and everyone took their bracing positions. Shirley was explaining the positions on the PA. It seemed like forever, and then the captain's voice: "I've got the runway in sight. I'm gonna shoot for the runway."

We had a very smooth touchdown and came to a stop.

I then heard a loud pop, and we [were] then hosed down with foam by the fire trucks.

Cheers of disbelief roared through the cabin!

FLIGHT ATTENDANT LINDA PALMER SCOPETTA

. . . We were coming in very close to the water, and we still did not know when ditching would come. I talked to [some passengers] while we sat braced for impact and told them to remain calm and that everything would be OK.

Other passengers were sighting land, and I knew the water had passed under us. I then very quickly briefed [nearby passengers] for an emergency land evacuation, just in case.

The cockpit door opened. I vaguely overheard the second officer tell Shirley that the landing would be normal. I still could not believe this and was still mentally prepared for evacuation.

After landing, everyone applauded and took in a breath of relief. I was in the cockpit when the first officer gave his report to the [passengers]. I could barely hear the actual report [because of the shouting and applause], but only vaguely heard about the two engines out, losing power, and that we would be towed in from that point. I knew that I would get the rest of the report later, and it was now that I was going to give my thanks to God.

The people . . . thanked me and hugged me and had hundreds of questions. One lady fainted, but she was calmed and was later very coherent and thankful.

6. PORTLAND, OREGON

December 28, 1978

United Airlines Flight 173

United Airlines flight 173 crashed 6.1 miles southeast of the Portland International Airport. About an hour prior to the crash, the flight had requested a delay from Portland Approach Control. The flight had encountered a problem with its main landing gear during initial descent. Portland Approach Control provided headings. Approximately one minute prior to the crash, while flight 173 was on a visual approach to runway 28L, the cockpit crew declared that it was "going down." The cockpit radioed to the Portland Tower, "Ah, Portland Tower, United . . . ah . . . one seven three heavy, MAYDAY . . . MAYDAY . . . ah . . . we're . . . the engines are flaming out, we're going down, we're not [going] to be able to make the airport." Of the 189 passengers, 8 died. Two crew members also died. The aircraft, a DC-8-61, was destroyed.

On that afternoon, the Douglas DC-8-61 of flight 173 required 46,700 pounds of fuel, and the captain saw no reason to worry that there was enough for them to reach their Portland destination. He could not have anticipated the landing-gear problem that arose as the flight was making its approach to Portland's runway 28. For the next twenty-three minutes, flight 173 flew a holding pattern while the flight crew discussed emergency procedures. The captain called United Airlines' Maintenance Control Center in San Francisco to get technical help analyzing their problem. He said, "I'm not going to hurry the [flight attendants]. We got about 175 people [there were in fact 189 aboard] on board and we . . . want to . . . take our time and get everybody ready and then we'll go [in for an emergency landing]. It's clear as a bell and no problem." The holding pattern kept flight 173 at five thousand feet and never more than twenty miles from the airport.

COCKPIT VOICE RECORDER

CAPT [to flight attendant]: How you doing?

FA [in the cockpit]: We're ready for your announcement [to passengers]. Do you have the signal for [when passengers should assume] protective position? That's the only thing I need from you right now.

CAPT: Okay. Ah . . . What would you do? Have you got any suggestions about when to brace? Want to do it on the PA?

FA: I . . . I'll be honest with you. I've never had one of these [emergencies] before. My first, you know. . . .

CAPT: All right. What we'll do is we'll have Frostie [second officer, flight engineer], oh, about a couple minutes before touchdown signal for brace pattern.

ATC: 173 heavy, turn left, heading two two zero.

FO: Left two two zero.

FA: Okay.

CAPT: And then, ah . . .

FA: And if you don't want us to evacuate, what are you gonna say?

CAPT: We'll either use the PA or we'll stand in the door and holler.

FA: Okay, one or the other. Ah, we're reseating passengers right now, and all the cabin lights are full up.

The flight attendant leaves the cockpit, and the cockpit crew converses over several minutes, giving voice to their fears.

FO: How much fuel we got, Frostie?

SO: Five thousand [pounds].

AN OFF-DUTY CAPTAIN WHO COMES INTO THE COCKPIT: Less than three weeks, three weeks to retirement. You better get me outta here [safely].

CAPT: Thing to remember is, don't worry.

OFF-DUTY CAPTAIN: What?

CAPT: Thing to remember is, don't worry.

FLIGHT ATTENDANT MARTHA FRALICK, AGE THIRTY-THREE, D JUMP SEAT

Approximately one half hour before landing a terrific jolt shook the entire airplane. I was standing in the aft buffet and was nearly thrown off my feet. I immediately sat down on my jump seat and called the cockpit, asking that some explanation be made to our obviously alarmed passengers.

I do not know to whom I spoke, but he replied, "Yes, in a minute."

FLIGHT ATTENDANT DIANNE WOODS

There was a crashing sound. It seemed to have come from underneath the airplane. A hitting-type force accompanied the crashing sound, and [Flight Attendant Martha] Marty [Fralick] and I both went to the aft jump seat, and I asked [Marty] to call the cockpit to find out what had happened. I heard her say to the person on the phone that they should explain what that noise was.

Soon after that, the captain gave an announcement. Just prior to his announcement we started looking at our manuals "just in case." A few minutes after that announcement, the cockpit called back. Marty answered. She said that the captain wanted to see us. We went forward. I stopped at the forward galley where Nancy and Sandy were. Marty went up [to the cockpit] with Joan [Wheeler]. They both seemed to come back right away. Joan said that the captain wanted us to prepare for an emergency landing. At that time the captain was on the PA again because someone had told him that the passengers could not hear his first announcement. He finished the announcement by telling the passengers that we would be preparing for an emergency landing and assured them of our excellent training and capability.

Marty and I went back again to review our specific duties. I went up to first class to tell Joan what Marty and I were about to do. She was preparing to do the announcement and I showed her where it was [in the manual]. Marty and I had just gone over it. Everyone started preparing the passengers. The next time I talked to Joan, she had walked to the rear of the airplane, checking to see if things had been done. At that time, we had already talked to the passengers about the brace position. In the captain's announcement he had informed the passengers that we had about twenty minutes before touchdown. He further said that he would be back [on the PA] about five minutes before touchdown [to] give them the brace position signal. We were all ready and had briefed, assigned and reassigned seats.

FLIGHT ATTENDANT MARTHA FRALICK, AGE THIRTY-THREE, D JUMP SEAT

Several passengers had their safety cards out and were looking at them. [Flight Attendant] Dianne [Woods] and I also began discussing the possibility of an emergency of some sort when [Flight Attendant] Joan Wheeler came to the rear and suggested that even though nothing official had yet been communicated to her by the cockpit, we might take out our manuals and review emergency procedures.

While I was getting my manual I heard parts of an announcement from the cockpit saying the jolt had something to do with the extension of the landing gear and we would be circling the field in order to check it out. A few moments later, this announcement was repeated, as it had not come through clearly the first time. I heard the crew-to-crew call bell and answered the phone. A male crew member's voice said, "Come to the cockpit."

I told Dianne we were wanted in the cockpit and the two of us went forward. When we reached the first coach buffet, [Flight Attendants] Nancy [King] and Sandy [Bass] were there. A decision was made that since two flight attendants should be in the cabin at all times, Dianne and I would be briefed first, followed by Nancy and Sandy.

As I reached the cockpit door and knocked it was opened by Joan. I said, "We were called to the cockpit." Joan's reply, as she pushed me backward from the cockpit door, was, "They meant me. We are to prepare for an emergency."

Joan did not say anything further so we proceeded to the rear and began our emergency preparations: reseating and briefing "helper" passengers, instructing passengers in the brace position, handing out pillows and blankets for padding, removing and storing baggage in the closets, and helping parents prepare their small children. All passengers with whom I had contact were cooperative, receptive to my instructions, and agreeable to reseating when necessary. As we started our preparation, the crew announced we would be taking this precaution in case the landing was other than normal.

Then Joan made her announcement, and I stood at my briefing/demo area and held up the emergency card and proceeded to elaborate on the brace position several rows at a time. I had several children put on their coats as I was sure that they would waste valuable evacuation time trying to put them on later. When I had completed all preparations and rechecked my "helpers" and made sure everyone understood their instructions, I proceeded to my jump seat and sat down. This action was purely instinctive. I recall that as I put on my seat belt and harness, I caught a fleeting glimpse of the No Smoking sign, but had no more than noticed it when the cabin went entirely black, except for the lighted emergency exit signs.

FEMALE, AGE FIFTY-ONE, SEAT 8C (THIRD-PERSON NARRATIVE)

She and her two daughters, ages fifteen and nineteen, boarded in Denver. Their flight was uneventful until the landing gear was lowered, making the "worst" sound, like metal "protesting." This caused the airplane to shake and almost knocked over a flight attendant. As a result, she told her daughter that there was a problem with the landing gear.

About ten minutes thereafter, a garbled announcement which she could not understand was made over the PA system. Her daughter, who had a headset on listening to music, said the announcement was about some problem with the landing gear. Several flight attendants were nearby and apparently had not heard the announcement. This passenger informed them and one flight attendant went forward to talk to the captain. She recalled that a cockpit crew member went toward the rear of the aircraft with a flashlight. An announcement was made that the gear was down, but they were not sure if it was locked, that they would circle and prepare for an abnormal landing and that passengers should listen to the flight attendants. She noted the flight attendants hurriedly checking their manuals and one flight attendant made what she interpreted as a very nervous announcement of emergency procedures.

Thereafter she was moved forward to 4C. She recalled that preparations were well made, that she understood how she should brace herself and that passengers were told they would have a final warning.

MALE, AGE THIRTY, SEAT 5F (THIRD-PERSON NARRATIVE)

The lowered landing gear [made] an abnormally loud and jolting noise, like something grabbing the plane and shaking it intensely, [and] from that moment he thought there was a problem.

He recalled that a few minutes later the pilot announced something abnormal with an indication of landing-gear problem, and that a check would be made [of the landing gear]. He recalled another announcement to the effect [that there was a problem with the landing gear] was made about five minutes later because the PA system was not working well. He estimated that the plane circled about an hour from the time of the gear problem to impact.

He recalled that the flight attendants checked their manuals and went through the aircraft preparing passengers for an abnormal landing with the possibility of an ensuing emergency evacuation, and that in his estimation there was no more they could do.

FLIGHT ATTENDANT NANCY KING

I need to explain a confusing sequence of events. After the original jolt in the air, the second officer came back into the coach cabin to look out the windows for an indication the landing gear was down. On his way back to the cockpit, [Flight Attendant] Joan [Wheeler] and I were standing by the first-class galley. The second officer stopped and quietly said, "You don't have a lot of time left. We can't circle all evening because we don't have a lot of fuel left."

I originally thought we landed approximately ten minutes after he made that comment. It is very confusing to me exactly when I spoke with the second officer. It must have been just before we started our preparation. I originally thought it was toward the end of our preparation, but I believe the second officer came into the cabin once to look out the windows.

I started briefing passengers in row 4, as to brace position and available exits. Joan had made the announcement asking for any company employees, policemen, firemen, or military personnel to identify themselves to the flight attendants. The man sitting in seat 4C identified himself to me as a policeman, so I moved him to 8C to assist at the forward galley jet escape exit. The lady originally sitting in 8C was moved to 4C. A gentleman in 7C identified himself as having retired from the Coast Guard. He stayed in his original seat to help assist at the coach galley door exit.

A major concern of mine was a couple in seats 4E and F, traveling with their six-month-old child. The lady did not speak English, so her husband translated my instructions of padding the baby on the floor and for both of them to hold on to the baby when they took their brace positions of grabbing ankles. A couple in seats 4A and B had a toddler boy. I took as many pillows, blankets, and coats as possible to pad the children on the floor. When I entered first class, the passengers had been briefed by Joan and seemed confident. I reviewed with them the exits and got more blankets and pillows for the children. I made sure a little girl who was seated in seat 2C was well packed with pillows and blankets between her and the seat.

I remember the brace signal was never quite clear to me, but at some point Joan had told me that the captain would either come over the public-address system or yell from the cockpit as to when to brace. I do not remember being asked or telling Joan personally that my preparation was complete. I had reviewed, again, [the emergency procedures] with the passengers. The babies were down on the floor, and I walked into the first-class galley. I looked out the galley window and felt [that] we were unusually low, seeing lots of streetlights, but I do not remember actually seeing the streets. At that point looking out the windows two times, I looked toward the passenger across from the galley and more or less talking to myself, I said, "Something is strange. I think we are landing." My memory is somewhat blank at that point.

COCKPIT VOICE RECORDER

OFF-DUTY CAPT: If I might make a suggestion. You should put your coats on, both for your protection and so you'll be noticed, so they'll [the rescuers and passengers] know who you are.

CAPT: Oh, that's okay.

OFF-DUTY CAPT: But if it gets, if it gets hot, it sure is nice not to have bare arms.

CAPT: Yeah, and if anything goes wrong, you just charge back there and get your ass off [the airplane], okay?

OFF-DUTY CAPT: Yeah, I told the gal to put me where she wants me. I think she wants me at a wing exit.

CAPT: Okay, fine. Thank you.

FO: What's the fuel now, Buddy?

CAPT: Five [thousand pounds].

The captain decides to remain in the holding pattern another fifteen minutes to burn off fuel, from five thousand pounds down to slightly more than three thousand pounds.

FO: Maintenance have anything to say?

SO: He says, "I think you guys have done everything you can," and I said we're reluctant to recycle the landing gear for fear something is bent or broken, and we won't be able to get [the gear] down.

FO: I agree.

CAPT: Ah, call the ramp and give 'em our passenger count. . . . Tell 'em we'll land with about four thousand pounds of fuel and tell them to give that to the fire department. I want mechanics to check the airplane after we stop . . . before we taxi [to the gate].

SO: Yes, sir. [Then on the radio to Portland]: Flight 173, we will be landing ah . . . in . . . ah, a little bit, and the information I'd like for you to pass on to the fire department for us [is that] we have souls on board one seven two—one hundred and seventy-two, plus five ba . . . ah, lap . . . ah, children.

PORTLAND: Okay, thank you.

SO: That would be five infants, that's one seven two, plus five infants . . . and we'll be landing with . . . about three thousand pounds fuel and . . . ah, requesting as soon as we stop that mechanics meet the airplane for an inspection prior to taxiing further. . . .

SO: Wind is three four zero at eight.

CAPT: Okay. . . . You want to be sure the flight bags and all that stuff is stowed and fastened down.

FO: How much fuel you got now?

SO: Four . . . four thousand . . . pounds.

FO: Okay.

CAPT [to second officer]: You might . . . you might just take a walk back through the cabin and kind of see how things are going, okay? I don't want to . . . I don't want to hurry 'em, but I'd like to do it [attempt emergency landing] in another . . . oh, ten minutes or so.

The second officer leaves the cockpit.

FO: If we do indeed have to evacuate, assuming that none of us are incapacitated, you're going to take care of the [engine] shutdown, right? Parking brakes, spoilers and flaps, fuel shut-off levels, fire handles, battery switch and all that?

CAPT: You just haul your ass back there and do whatever needs doing. I think that [Flight Attendant] Jones is a pretty level gal . . . and sounds like she knows what she is doing, and Duke [the off-duty captain] has been around for a while. I'm sure Duke will help out.

The second officer returns to the cockpit.

CAPT: How are the people?

SO: Well, they're pretty calm and cool and, ah . . . some of 'em are obviously nervous . . . ah, but for the most part, they're taking it in stride. I stopped and reassured a couple of them. They seemed a little bit more anxious than some of the others.

CAPT: Okay, well, about two minutes before landing—that will be about four miles out [from the end of the runway]—just pick up the mike and say, "Assume the brace position."

Several more minutes pass while the crew discusses fuel levels and preparations for an emergency landing.

CAPT [to flight attendant in door]: Okay, we're going to go in [to land now]. We should be landing in about five minutes.

FO: I think you just lost number-four engine, Buddy. You . . . .

FA: Okay, I'll make the five-minute announcement.

FO: Better get some cross feeds open [to engine number four], or something. We're losing an engine.

CAPT: Why?

FO: Fuel! Open the cross feeds, man.

CAPT: Open the cross feeds or something.

SO: Showing fumes [in the tanks].

CAPT: Showing a thousand [pounds of fuel] or better.

FO: I don't think [that much] is in there.

SO: Showing three thousand, isn't it?

CAPT: Okay, it's . . . it's . . .

FO: . . . a flameout!

CAPT [to Portland ATC]: 173 would like clearance for an approach into 28 Left. Now!

SO: We're going to lose number three [engine] in a minute, too.

CAPT: Well . . .

SO: It's [the fuel gauge is] showing zero.

CAPT: You got three thousand pounds. . . . You got to. . . .

SO: . . . [We're supposed to have] five thousand pounds of fuel in there, Buddy, but we lost it.

CAPT: All right.

SO: Are you getting [reignition of the engine] back?

FO: No! You got that cross feed open?

SO: No. I haven't got it open. Which one . . . ?

CAPT: Open 'em both. God, get some fuel in there. Got some fuel pressure?

SO: Yes, sir.

CAPT [trying to start the fuel-starved engine]: Rotation. Now she's coming. Okay. Watch [engines] one and two. We're showing down to zero [on fuel] or a thousand [pounds].

SO: Yeah.

FO: Still not getting it [the engine started].

CAPT: Well, open the cross feeds.

SO: All four?

CAPT: Yeah.

FO: All right now. It's coming. It's going to be a bastard on approach [to the airfield], though.

CAPT: Yeah. You gotta keep 'em running, Frostie.

SO: Yes, sir.

FO: Get this mother on the ground.

SO: Yeah, it's not showing very much more fuel. . . .

CAPT [to Portland Control]: 173 has got the field in sight now.

PORTLAND: Okay, 173. Maintain five thousand.

CAPT: Maintain five.

SO: We're down to one [thousand pounds] . . . Number-two tank is empty.

CAPT [to Portland]: 173 is going to turn toward the airport and come on in.

PORTLAND: Okay. Now, you want to do it on a visual. Is that what you want?

CAPT: Yeah.

PORTLAND: Okay, 173. Turn . . . ah, left, heading three six zero and verify you do have the airport in sight.
FO: We do have the airport in sight.
PORTLAND: 173 is cleared visual approach, runway 28 Left.
FO: Cleared visual, 28 Left.
CAPT: Yeah.

The sounds of the engines still operating being reduced in speed.

FO: You want the Instrument Landing System on, Buddy?
CAPT: Well . . .
FO: It's not going to do any good now.
CAPT: No, we'll get that damned warning thing if we do.

About a minute goes by.

CAPT: Ah, reset that circuit breaker momentarily. See if we get [landing] gear [indicator] lights. Yeah, the nose gear's down. About time to give that brace position [announcement to the passengers].
SO: Yes, sir.
CAPT [to Portland]: How far you show us from the field?
PORTLAND: Ah, I'd call it eighteen flying miles.
CAPT: All right.
SO: Boy, that fuel sure went to hell all of a sudden. I told you we had four [thousand pounds].

Another minute goes by.

CAPT: There's . . . ah, kind of an interstate highway–type thing along that bank on the river—in case we're short.
FO: Let's take the shortest route to the airport.

FLIGHT ATTENDANT SANDRA BASS

I was forward in coach about row 12 when I noticed the No Smoking/Fasten Seat Belt sign was on. I think it was dark then but it did not faze me then. I shouted, "Put your cigarettes out now." I noticed a female passenger smoking in the non-smoking section. She put out her cigarette. I all of a sudden noticed the cabin to be dim and extremely quiet. It was like a sixth sense. I felt it was time to sit. I went to 3R and secured myself in my jump seat. You could have heard a pin drop, it was so quiet. I don't think I fully realized we had lost power. But I knew we were very close to landing. I thought we were going into the airport. I had no feeling of whether we would be crashing or landing normally. I just knew we were close, and it was too quiet.

MALE, AGE THIRTY, WITNESS ON THE GROUND

We were out in the street about to get into the car when we saw the plane overhead. He was approximately one hundred fifty feet high, heading in a northeasterly direction. The plane was completely dark, no lights at all, and completely silent, no engine noise at all. I saw the landing gear in the down position.

FLIGHT ATTENDANT NANCY KING

I only remember Joan was standing there by seat 8C and I said, "I'm going to my jump seat."

Joan headed forward. . . . I had enough time to sit down, fasten my seat belt, and recheck to make sure it was as tight as possible. My brace position was "Feet flat on floor," and I crossed my hands holding on to my harness with my head down. Before I held on to my harness, the full cabin lights went out and the only lights visible were the Exit signs over the individual exits. It only seemed like a matter of seconds with the darkness and complete silence—like a glider.

Right before impact, I heard Joan yell, "Grab your ankles."

I screamed out, "Grab your ankles. Keep your head down until the plane comes to a complete stop!"

FLIGHT ATTENDANT MARTHA FRALICK, AGE THIRTY-THREE, D JUMP SEAT

There was absolute silence, and I remember wondering if it was always this quiet before a crash . . . no baby crying, no child's voice, no nervous coughs.

I saw bright Christmas lights through the passenger window in front of me, but it somehow did not connect that we were not landing at the airport.

COCKPIT VOICE RECORDER

SO: We just lost two engines. One and two.
FO: You got all the pumps and everything [working]?
PORTLAND: . . . You're about eight or niner flying miles from the airport.
SO: Yep.
PORTLAND: Have a good one.
CAPT: They're all going! We can't make it!
FO: We can't make anything!
CAPT: Okay, declare a Mayday.
FO: 173, Mayday! We're . . . the engines are flaming out. We're going down. We're not going to be able to make the airport!

FLIGHT ATTENDANT DIANNE WOODS

Everything was very quiet, and we seemed to be floating down. I heard someone say, "Grab your ankles. Hold your head down." And I did the same thing. In another moment we appeared to land with a ka-bonk! Ka-bonk!

FLIGHT ATTENDANT SANDRA BASS

I started shouting on impulse before impact. I never heard a cockpit "brace position" announcement. I immediately heard [the off-duty captain] shouting, "Grab your ankles." I was shouting, "Grab your ankles, keep your heads down," several times. I felt a forward and side thrust. I heard a loud sound of the aircraft being torn apart. I saw flashes of bright light, like electrical wiring arcs and blue-white flashing. I was grabbing my own ankles while shouting. The swerving and jolts were quick, not more than four or five seconds. Then it was dead still.

FLIGHT ATTENDANT MARTHA FRALICK, AGE THIRTY-THREE, D JUMP SEAT

Then came the initial impact, and [I was] caught somewhat unaware. I nonetheless began shouting, "Keep your head down," and "Grab ankles." My arms and legs were thrown out in front of me, and I saw things flying through the air, but could not identify what they were. There were three bumping, grinding, jarring, forward-throwing impacts. Some swerving occurred, but the major impetus was forward.

MALE, AGE THIRTY, WITNESS ON THE GROUND

A few seconds later, I saw a big flash of bright white light, then crashing scraping noises, then another yellow flash, then another bright white flash, and a loud crashing boom.

FEMALE, AGE FIFTY-THREE, SEAT 20A (THIRD-PERSON NARRATIVE)

She felt one impact and thought the landing was not so bad. Then the wall [of the aircraft] came in on the left side and hit her in the side, knocking her wind out [of her lungs].

FEMALE, AGE FIFTY-ONE, SEAT 8C (THIRD-PERSON NARRATIVE)

Within seconds of impact, she recalled, the light in the cabin went out, and she thought how strange that [the pilot] would land with the lights out. Then she heard voices screaming, "Grab ankles!" She noted some trees outside and impact with trees. She thought they had landed. Then a jolt and several jolts and the partition in front of her disappeared.

MALE, AGE THIRTY, SEAT 5F (THIRD-PERSON NARRATIVE)

He was unable to recall any variation in the lights. He was in the crossed-arms brace position when things got very quiet, and the plane was gliding. He thought they were landing at the airport and recalled some mild impact with trees and then a couple of very hard jolts.

FLIGHT ATTENDANT NANCY KING

. . . Suddenly we hit with a force so outrageously strong that I found my body tossed about inside my seat belt and my legs and arms swinging every which way about me.

I was trying to bring my arms around my head for protection from the objects flying past me [at] incredible speed. The sound of the wind I will never forget. The deep, strong whistling sound—the frightening speed and force—carrying with it soft-drink cans, paper and trays from the galley. The last object I remember was the metal cover to one of the slide packs from the galley. It appeared to come from behind, as the wind was blowing aft to forward, around the corner from my jump seat and landed with a fierce halt at my feet.

After three quick crashing bounces and swerving of the airplane we came to an abrupt halt. I began screaming, "Release your seat belts and get out!"

FLIGHT ATTENDANT MARTHA FRALICK, AGE THIRTY-THREE, D JUMP SEAT

Instants later, we stopped. The cabin was filled with a dry, choking dust. I turned to my control panel, signaled the cockpit with the rapid chimes, all the while thinking, "Why don't I hear these chimes," and wondering if I was hitting the wrong button, and [I] switched on the emergency lights.

At the same time, I was shouting, "Unfasten seat belts and get out!"

Passengers were immediately on their feet and moving toward the exits. I jumped up and after looking to see if there was fire, started to open my door. Over my shoulder I shouted to my "helper" passengers to check for fire before they opened their door. I got my door partially open when I heard a loud, ominous hissing. I began choking from a dry gaseous odor. I thought there must be some kind of gas leak and momentarily thought there might be an explosion. I paused, almost thinking I should try to close the door to keep out fire, but the hissing stopped and I proceeded to try and open my door.

I could not open my door more than a third of the way, when a male passenger behind me asked if I needed help. I replied, "Yes. I can't get my door open," and he stepped in front of me and forced it the rest of the way open.

FLIGHT ATTENDANT DIANNE WOODS

"Release your seat belts and get out!" was shouted.

Within seconds, it seemed, everyone was up. I all but pulled the handle off my jet escape door exit, but to no avail.

Immediately, I looked to the other exits. At this time my "helpers" were kicking at my exit door. Most of the passengers seemed to be already out of the airplane, except for those near my area. Soon these were all gone too, after I redirected them. I told them to get out!

Other than not being able to open my exit door, everything from up to where the seats buckled, about row 19 or 20, everything went extremely well. There was no screaming or confusion. Sandy and the passenger "helper" were [on the] forward side of where the seats were buckled. I could see the "helper" with the flashlight.

FLIGHT ATTENDANT SANDRA BASS

I shouted, "Release your seat belts and get out!"

I undid the PA phone cover and punched the cockpit button many times. I saw no fire. I tried to open [emergency exit door] 3R. It was jammed. Several passengers tried and kicked at it. It wouldn't open.

I could see a bluish haze, and I feel like I inhaled some fumes. I didn't sense them to be fire or smoke, though. I did not sense fire. But I can remember a burning sensation in my throat.

Three-R was not usable. So I directed the passengers who were all in the aisles by now that "there are exits in front and behind you. Come this way." I immediately hopped over the seats to 4R. It was already open, but the slide was crooked into some trees. I realized then we weren't at the airport. I asked a man what he could see. He said,

"About one hundred feet there is a clearing." I asked if he could get there. He said, "Yes." I said, "Go. Direct people away from the plane." He moved fast. People started moving quickly through the exit. I heard no screams. I saw people exiting through the back. Four-R was open.

FEMALE, AGE FIFTY-THREE, SEAT 20A (THIRD-PERSON NARRATIVE)

She thought the blow had knocked her out, and she wanted to relax, but a man with a flashlight was yelling for her to get out. She was irritated by this, but unfastened the lap belt, got up, and walked forward to an exit. As she went through the exit onto the wing, she put her hand on a rubber tire. A rescue man helped her off the wing and told her to go toward a blue flashing light. She detected a faint smell of fuel. She received rather severe contusions to the left chest wall, fracturing four ribs when a main landing gear struck and partly penetrated the left side of the aircraft where she was sitting. She also received contusions in the pelvic area consistent with the lap belt.

FEMALE, AGE FIFTY-ONE, SEAT 8C (THIRD-PERSON NARRATIVE)

The next thing she remembered was hanging from her seat almost face down—seat tilted forward—and being unable to move her right arm. She thought if she released her lap belt, she would fall to the ground.

A man passed her going aft and said something like, "I'm going to get out!" Another man came forward and grabbed her slacks at the waist to hold her secure while she released her lap belt. He then hoisted her up to the floor level and guided her aft in the direction of the exit. She walked aft and went out 2L exit after observing the "chute wasn't right." She climbed out over some debris and heard her daughters calling for her. In recalling the cabin during her evacuation, she said that there was a gray-type light inside, and she could see OK. No smell of fuel, only dust. She had to climb over a lot of debris in the aisle and recalls one large pile. This passenger had a fracture of both clavicles, fractures of the right arm at the elbow and wrist and numerous contusions and abrasions.

MALE, AGE THIRTY, SEAT 5F (THIRD-PERSON NARRATIVE)

The next thing he recalled was being in darkness, in rubble. It was dark. He could not see with his left eye. There was blood in his eye from a head laceration. He realized his daughter was next to him. He reached over and unbelted her lap belt, picked her up, and carried her away from the aircraft, hearing a command that he should walk to the right.

At this time, his leg almost collapsed [under] him and he realized that his shoes (laced Hush Puppies) were missing.

FLIGHT ATTENDANT NANCY KING

Immediately, the gentleman seated in 8C had opened the jet escape door in the first-class galley. It was a frightening shock to see and hear the hissing sound of the slide quickly inflating inside the galley and extending out into the aisle. I remember hearing screaming passengers at that point, and I was backing away with the fear the slide would slam me up against the wall. Within an instant the inflated slide touched seat 8C, and I remember the enormous relief when it stopped and the hissing sound was gone.

I then turned to my jet escape door to the left of my jump seat. It quickly opened with the slide inflating outside the airplane, but only went about one third of the way down, and the remaining portion of the slide appeared totally tangled in the trees and large branches. I held on to the inside of the airplane and leaned out so that I could see exactly where the passengers would be going. I distinctly remember the total confusion as to where we were?!?

FLIGHT ATTENDANT SANDRA BASS

I then looked forward to notice some passengers still forward of me. I went forward over the seats past the people in the aisle. I could see Martha behind me directing people. All but a few were out forward. I came to row 23 when I noticed [buckled] seats between 21 and 23. I saw a man sitting in a B seat stunned. I shouted at him and slapped him. He was dazed but alive, and I asked if he could walk. He groaned and said, "Yes." I asked, "Are you OK?" He said yes dopily. I lifted him out of the seat. He was holding on to me and we went forward. [The off-duty pilot] saw me and assisted in getting him off the airplane. I remember shouting at two more passengers who were moving slowly about row 19 to "get out now." They were OK and went out a window.

Martha yelled at me from the back to get the flashlights from the forward closet. I remember looking forward and seeing the big "hole." I yelled back, "There is no front closet. There is no front cabin." I then went forward.

I saw the slides employed inside the cabin at 1R and 2R. I crawled over them. I saw Nancy. She [was] OK. I didn't see Joan. All the passengers in Zone 2 rows 4 to 8 were out. I then went back through the coach cabin checking and shouting, "Is everyone out? Is anybody hurt?" I shouted over and over. I looked. I had my mini-flashlight out and the emergency lights were on. I could see enough to see no one else was on [board].

I then went forward and Nancy and I went out 2L. The slide was inflated only at the bottom. The deputy sheriff was there by then. Nancy and I went around to the front to help get people out.

FLIGHT ATTENDANT NANCY KING

The feeling that we were in the treetops frightened me because I was afraid of the distance below and how far down I was sending the passengers [down the slide]. I was terribly frightened by the odor in the air, which felt like breathing in and swallowing gas. I was conscious of slowing my breathing down, thinking the so-called gas could be toxic. Another fear of mine was the amount of time we had before fire broke out—thinking it would be any second. People were crawling over the seat backs and had to crawl over the inflated slide from the forward galley to get to the jet escape exit. I had to slow down the movement of people going out the exit because the people were toppling over one another outside on the slide. They looked like a lot of people trying to walk on a trampoline at the same time and falling on top of one another. I only had two passengers who hesitated to go out my exit. One was a small boy, maybe fourteen years of age, whose face was covered with blood. The other was a lady who thought her shoulder was broken. I was helping her crawl over the inflated slide in the aisle. She said to me, "I can't go any farther. I think my shoulder is broken." I was sympathetic and told her I was sorry she was hurt and immediately walked her to the exit and firmly said, "You must get out!" I yelled to passengers outside on the slide [that] I had an injured lady and to help her down. That lady was the last passenger I remember to go out my exit. I would say approximately ten passengers went out my jet escape exit.

FLIGHT ATTENDANT MARTHA FRALICK, AGE THIRTY-THREE, D JUMP SEAT

I looked down, and the slide was askew but inflated. People were already moving slowly but steadily out of [emergency exits] 5R and a few started out 5L. When I perceived this section of the airplane was being evacuated, I began to work my way forward to see if any help was needed further up.

People were shaken and upset, but there was no . . . panic. By the time I reached [emergency exits] 3R and 3L, the airplane was almost empty. I could see the coach cabin clearly by the light of the emergency lights and saw that, with the exception of about two rows of buckled seats, the fuselage was intact. There was no smoke or fire, and I could also see a portion of the passageway adjacent to the first coach galley, which also was intact. I had seen Sandy go forward and felt my next responsibility was to get out of the airplane, collect my passengers, and see who was injured. I went to the rear of the airplane and got the megaphone and first-aid kit. At this time Dianne was in the rear closet looking for the flashlights. We both went forward to the 3R and 3L exits. I went out 3R. By this time there were no people left on the airplane. And Dianne went out 3L.

One of our "helper" passengers was outside 3R and helped me, as I climbed over the roof of a collapsed house and over and down several fallen tree trunks. I looked around and saw lights to my right. These turned out to be streetlights on Burnside [Street]. When I reached the road I used the megaphone to call my passengers toward me. Almost immediately a county sheriff's deputy appeared at my elbow and told me to direct everyone to the "bright light," which was the light at the intersection of Burnside and 157th [streets]. This was ultimately where the rescue and organization efforts centered. I continued toward this intersection, where I encountered Dianne. After determining there were sufficient rescue workers and sheriff's officers to take over organizing and caring for the injured, Dianne and I decided, as we had not seen Sandy, Nancy or Joan, we should go to the front of the plane, see where they were, and if they needed help. A sheriff's deputy took us through the trees and backyards to the front of the airplane. We were astounded at what we saw: There was nothing left of the front of the airplane but debris and bodies. The torn front end of the fuselage was swarming with silver-suited firemen. Amid the debris on the ground I could only see one set of passenger seats still intact with some people in it.

FLIGHT ATTENDANT NANCY KING

During the time passengers were going out my jet escape I had yelled to Joan twice. I received no response. Immediately after the last visible passenger went out my available exit . . . I headed forward toward the first-class cabin. . . . To my complete surprise, shock, and horror, I discovered total debris with injured and dead passengers. Because of the total darkness in the airplane, other than the lights from the exit sign, I had no idea there was nothing left of the first-class cabin and the cockpit.

At that point in time my flashlight became my "right arm," so to speak. The policeman's daughter—I believe she was thirteen years of age—who was seated in seat 4D was nowhere to be found. The policeman kept saying, "She's gone. She's gone." With my flashlight I checked all around the remaining seat rows in my cabin to make sure no one was left and looking for the girl at the same time. It was so dark that it was difficult to tell what was around me. I knew it was a very large opening [and I felt that] I was in an enormous cave. Injured passengers were trying to make their way down a piled mound of debris. I was in my uniform skirt and constantly aware of the sharp torn pieces of metal everywhere. Along with passengers I was crawling down piled wreckage, hoping for the ground somewhere. In the wreckage outside I continued my search for the policeman's daughter with my flashlight. I was picking up pieces of wreckage and the horror of blood-covered bodies was overwhelming.

The correct sequence of events following [that] are all confusing to me. I located one of our pilots, who was facedown in the middle of the debris and only visible from the shoulders up. I vaguely remember someone telling me he was dead. I discovered the foreign gentleman who had been seated in seat 4F. He was obviously injured and sitting on the ground in front of all the debris. There were two passenger seats close to him that were intact but apparently thrown clear of the wreckage. With the help of some man I picked up the foreign man and placed him in a passenger seat, so he was facing the wreckage. The foreign man was covered with blood, shocked and crying out for his baby. I assured him I would try to find his baby. Emergency help had arrived. Lights were glaring, and people were scrambling everywhere. I saw a small hand of a child showing in the debris and helped to lift wreckage where we found a small girl, approximately three years of age, buried beneath.

FLIGHT ATTENDANT MARTHA FRALICK, AGE THIRTY-THREE, D JUMP SEAT

I looked for, but could not see, any of the other flight attendants and was approached by a fireman who asked what I wanted. I informed him [that] I was a flight attendant off this flight and had come to help. He replied [that] they had "the only survivor," pointed to the man in the passenger seat and told me to "get out of here." I looked around again, and seeing I could do nothing, called to Dianne, and we left that immediate area.

After returning to the intersection, I was approached by a young girl who asked if I could come look at her mother. She led me to a house where I discovered a living room full of injured people. I had taken an emergency rescue person from the intersection with me, and he started attending to the woman. I then left the house, taking an injured but ambulatory woman with me to the intersection. I returned to the house several times, taking injured out and directing others to the intersection. Finally, when I felt I had done all I could and all of the people I could find were attended to, I encountered Dianne. We were directed by a sheriff's deputy to a church a block away for transportation to the airport.

FLIGHT ATTENDANT DIANNE WOODS

Because everything had gone so smoothly in the back and I could not see the front of the airplane because of the trees and dark, I thought everything had gone just as smoothly in the front.

Marty and I were outside trying to get our people together and looking for Sandy and Nancy, but we never saw them come 'round. It was at least twenty-five minutes before we walked around to the front of the airplane. I was appalled. It seemed like two completely different accidents, one that went smoothly where no one was hurt and one tragic where people were bleeding and dying.

FLIGHT ATTENDANT SANDRA BASS

I remember lots of large flashlights. We could see people [were] hurt, but alive. We could see the disintegrated cabin area in a big heap. I could not find the cockpit but I believe I saw a dead pilot. At that time I assumed they were all dead. I looked around to find Joan. There were bright lights in the main forward crash area. It wasn't until later I realized some were camera lights.

I thought I saw a rust [-colored] uniform. I went over to a clearing away from a pile [of seats] under a tree. It was Joan. She was lying on her back, limp, and I think, dead. I shouted for help. I checked for [her] pulse and tried to dilate her eyes. I must have known she was dead. I walked over to Nancy and told her [that] I thought Joan was dead. Nancy heard crying. She said she saw an arm move under the pile of seats. [The off-duty pilot] and [I] immediately went over to the area. I remember tugging to release some coach seats off the pile. [The off-duty pilot] was lifting away fuselage. A rescue worker freed a small female child who was alive.

Help was all over by then. I could see dead people. I went back over to where Joan was lying. I checked her again. I once again shouted for help. A rescue worker came over and verified she was gone. He said he would get a paramedic over, get an EMT and pronounce her dead.

I went over to Nancy, who was reassuring an injured female passenger. The medics were at her aid by then. The Yugoslavian man was being aided. That's all the life I could see.

I knew we could do no more. I asked a rescue worker where everyone and other flight attendants had assembled. He said, "The church." I asked if we needed blankets, pillows, etc. He said they might help. It was cold out. Nancy and I went back into the craft to gather blankets and go to the church.

7. DENVER, COLORADO

November 15, 1987

Continental Airlines Flight 1713

Continental Airlines flight 1713 crashed during an attempted takeoff on Runway 35L at Stapleton International Airport, at approximately 2:16 P.M. in snowy, freezing weather. The left wing struck the ground initially and the aircraft rolled and inverted and broke up. There were seventy-seven passengers, including two infants, three cabin attendants, and two flight crew members on board. Both pilots, a cabin attendant, and twenty-five passengers, including one infant, died. The aircraft was destroyed.

COCKPIT VOICE RECORDER

Sound of deicing spray outside on the fuselage.

CAPT: It's like going through a car wash.

Sound of laughter. Then, sound of a knock on the cockpit door.

CAPT: Come in.
FA: Check this out. It came in the door. The deicing crud came through the door.
FO: Tight seal.
CAPT: Around the edge.
FA: Won't that depressurize [us]?
CAPT: No. We're not pressurized.

FO: Yeah, there's no pressure on the seal now. When we pressurize, the seal inflates.

FA: Oh, it will?

FO: See? That's why it came in.

The crew starts the engines after going through a before-start checklist, then it commences an after-start checklist. Then they begin to taxi out to the runway for takeoff.

CAPT: We're not going to get much slop [ice and snow] between here and the end [of the taxiway], so why don't you go ahead and get flaps down.

FO: Okay, flaps ten. Taxi check complete.

CAPT: Thank you. Ah, if we got to do this for seven landings, I tend to lose my enthusiasm.

FO: Yeah.

CAPT: Tired. . . .

FO: Yeah, no kidding.

CAPT: This stuff [snow and ice] may not go anywhere. It may hang in here for a couple of days.

FO: Uh-huh.

CAPT: A stationary low like this.

A minute goes by as the flight waits in line on the taxiway for takeoff.

CAPT: You might tell him that we're, ah, number one [for takeoff] here on the north side.

The FO radios the tower to tell them that their flight is "number one DC-9 for Continental."

CAPT: It didn't impress him at all.

FO: Apparently not.

The tower appears to be confused about which plane is which.

CAPT: Okay, I think we got 'em straightened out.

Sounds of laughter. For the next five minutes the crew edges up in line, conducting "nonpertinent" conversation.

TOWER: Continental 1713, taxi into [takeoff] position and hold.
FO: Position and hold for 1713.
FO: Ladies and gentlemen, we have been cleared to taxi into position and hold on the runway. We will be airborne very shortly. Flight attendants, prepare the cabin for departure and take your seats.
TOWER: For whoever that was that asked company reports, there is a little crud on the runway. I don't know how to define that [in other words].
FO: A little crud. Speeds—139, 145, 159 . . . .Departure is [frequency 123.8], and I got it over here too.
CAPT: Don't slide. You might tell him we're in position [for takeoff].
FO: Okay.
CAPT: Okay, Red Rover. . . .

Sounds of laughter.

FO: Bend over and bark like a dog.

Sound of whistling.

CAPT: Got the brakes on. You got the airplane [for takeoff].
FO: Okay.
CAPT: I got the radio. Run 'em [the engines] up a little bit before you release the brakes and let them stabilize.
FO: Okay.
TOWER: Continental 1713, runway 35L, cleared for takeoff. Wind is three six zero at one four. Runway visual range, two thousand [feet].
CAPT: Cleared for takeoff, Continental 1713.

Increasing engine sounds.

CAPT: Light on. Okay, power's at . . . Left and right . . . We got ninety-five, ninety-three [knots].

Eleven seconds go by.

CAPT: There's a hundred knots, looking for one thirty-nine. . . . Vee one. Rotate. Positive rate [of climb].

Sounds of engine compressor stall. Sound of bang. Sound of impact. End of recording.

MALE, SEAT UNKNOWN

The aircraft was crowded. . . . I only noticed the man who had to get up to allow me to take my seat, which was over the starboard wing, the trailing portion of the wing. And [I] really had no conversation with anyone. As soon as the aircraft was loaded, the captain announced we [were going] to the deicing area. And then his words, I think, [were] "takeoff immediately." We did, of course, taxi out to that [deicing] area. I was interested in that, because I had never been in a deicing situation before. And I could hear the machinery running over the fuselage of the aircraft as well as watch it come out over the starboard wing.

Then it seemed quite a long period of time. . . . We taxied to the runway, and during that time I do not recall any conversation with anyone. The stewardesses had given their instructions. And like all passengers or many passengers, I had not listened. But I had flown [on that type of] aircraft a number of times, so I was comfortable knowing where the exits were and all.

. . . When we reached the runway, the announcement was made that we would begin takeoff. I had noticed the stewardesses or the cabin crew come through the cabin. I did not notice anyone from the cockpit come back to look. At one point, preparatory to the takeoff, I remember looking at the wing, wondering, "Is that ice?" But I certainly did not see any clots of ice or snow or anything of that sort. And I rather decided it was probably only wet.

The roll down the runway at takeoff I thought was normal for Stapleton. It always seems to me it takes forever to get airborne here. It was the usual long roll. I did not feel it was unusual.

FLIGHT ATTENDANT KELLY ENGLEHART

I remember the captain saying, telling the flight attendants to take their seats for takeoff. And when they say that, we take off almost immediately [afterward]. And we started our roll on the runway. And I remember thinking . . . that the engines did not seem at full power. It was just a passing thought.

MALE, SEAT UNKNOWN

Now at this point . . . my perception was that, in fact, I was watching the ground [and that] the ground did drop away. I can't see [the ground] when things began to happen, but I think it could have been anywhere from ten to thirty to fifty feet [off the ground]. The aircraft lurched violently, dropped the wing rather on the right or the starboard side—the side I was seated on. My estimate would be twenty or thirty degrees. And there was an audible, almost a palpable, thump which I could equate to the normal wheel retraction when an aircraft, a [Boeing] 727 at any rate, takes off. Not a scraping noise, but a thump. Then my analogy is the same as if one loses a wheel off the end of the highway and violently overcorrects. It seemed that the aircraft came back suddenly to the left side. And, of course, at this time a lot of things were happening at once.

It was clear to me, I think, and probably everyone when it made the first dip, that we had a big trouble.

FLIGHT ATTENDANT KELLY ENGLEHART

And then we nosed up and then as soon as the wheels, the back wheels, lifted up, we . . . just went like this to the left side. And it made [another flight attendant sitting beside her] and [me] jump, because it was so soon after liftoff that the airplane went to the left like that.

And as soon as it came back down to the right, then it really went to the right. So much so that we felt the wing must have hit, because then it just ran into the ground.

Then I heard like three large pops and then a huge explosion. And I could see a big orange fireball at the front of the cabin just moving down [the aisle]. And at that time I just shut my eyes because I thought it was over. And I was just waiting for [the fireball] to get me. I think it happened so fast the next thing we knew we were upside down. And it was over, and there was no fire.

MALE, SEAT UNKNOWN

And I went into a crash position, a fetal position, my hands up over my head. There was no time for any announcements or anything, certainly. And I suppose there were some shouts or screams, but I really don't recall.

I was aware of a flash of light, which I remember as an electric light, rather than a fireball. And I did not see a fireball. As the aircraft came back to the left, it just kept going in that direction. I had the thoughts that Ms. Englehart [the flight attendant] has already described . . . [that] I am going to die. And a number of other thoughts go through your mind.

And then the next thing I knew, there was this sensation of impact, dust in the air, and a lot of noise.

My first thought was that we were still falling, because I could hear a whining noise that I think in retrospect must have been the turbine slowing down. We obviously still could not be falling.

At that point, when the noise subsided, I decided, "My God, I survived an air crash." And equally at the same time I was sure that I was going to burn. And I thought, "Well, I have got to get out of here," [but] there was no way I could move. I was compressed in that fetal position, if you will, still strapped in my seat. And subsequently, [I] decided [that] I must have been head-down probably thirty or forty degrees. It was as though there was just a steady firm pressure all over my body from the top of my shoulders to the tip of my toes from the back. My head rested against my left arm, which was up over my head. My right arm was free. And I had a little bit of discomfort in a finger on my right hand, which subsequently proved to have a break in it. My glasses were still on, partially. I had no bleeding or anything. And I was able to contract muscles and all. I tried to move the weight to see if I couldn't get up, but I couldn't do it. And I was sure I was going to burn.

FLIGHT ATTENDANT KELLY ENGLEHART

We were disoriented, so it was hard to even really—I don't think I ever realized at the time that we actually [were] upside down. But we automatically, as soon as we stopped, undid our belt and I rolled out of my seat. Once I got my bearings and I stood up, there [was] a door that was at a thirty-degree angle in front of me, like this. And I jumped over it to the other side. And I realized it was the cabin that I was in at that point. And then I jumped back over with [the flight attendant] and we were fumbling for the door trying to figure out—I don't think I knew that plane was upside down at that point.

And we were struggling with the door, and we could not get it open. So I jumped back over the door. This was a [lavatory] door that had broken off. It was wedged in between the two bathrooms at an angle. So I jumped back over the door and checked on the passengers. And everybody back in my section was—they were alive! It was amazing. And nobody was really seriously injured. And nobody was pinned. So I got two guys [who] were sitting in the back two seats and they jumped over the door and went to help [the other flight attendant] open the door. And at that point, I was sitting and all of a sudden I . . . checked on all the passengers to see if they were OK. The little six-week-old baby was just screaming, and everybody was so glad to hear that baby cry, because we knew the baby was OK.

MALE, SEAT UNKNOWN

Time passes, and that fear [of burning] subsided.

I smelled no smoke and heard no screams or noises suggesting fire. And as time went on, I became a little bit optimistic [that] I might make it through there. Please understand at that point, during the whole thing, I lost all concept of time. I was conscious all the time, but time did not mean a thing. And I think the reason . . . I was able to get through that, [I was able to] disassociate from where I was. If I thought much about the predicament, it was depressing to say the least. But if I could project to another place, it was much better.

Eventually I heard someone outside shout, "Relax, folks. We have got it foamed. Everything is all right." That was reassuring. A little bit later someone came by and said, "Is there anyone in there?" And a number of us shouted, "Yes." He said, "Will you count off?" And I think all of us shouted "one" at the same time. And I think there were probably five or six people in that area that counted [off].

I was aware of the man next to me. . . . He complained of pressure on his back. . . . The man who must have been in front of me, [his] head was kind of in my abdominal area but not pushing against me. And he was moving and talking. We were both aware of a fresh air supply and could get a little bit of light. We could not see out, but at least there was fresh air. We did not talk, or talked very little. There was a little bit of encouragement among the passengers. There was really no panic. There was some cries of pain. And there were some prayers, I think probably. But really, it was remarkably quiet, I think, in retrospect.

FLIGHT ATTENDANT KELLY ENGLEHART

We were trying to get [a door] open, which was underneath me, to get my emergency equipment. There were five extinguishers there. There is a first-aid kit. I couldn't get the extinguishers out then. So I was pulling and pulling and pulling, trying to get the first-aid kit out. And the first-aid kit on the DC-9 is stuck in a metal casing inside the overhead bin. And because it was in this casing that it fits perfectly into, that casing was jammed, so I could not even get the first-aid kit out.

So I grabbed the flashlights, so the people could see in the cabin. And it took me about twenty or thirty minutes. It seemed like that to me anyway, before the door was finally opened.

We got the exit door open, and we sent the couple with the little baby out first. Then we sent a little girl. And then the mother went. And I noticed that everybody was crawling over me to get out, so I crawled out after them. And I got everybody in a huddle, kind of to the left side of the aircraft. It was kind of past a fire truck. We all got in a huddle and put our arms around each other. We were kind of hugging each other, because we could not believe we were alive. It was . . . like, you know—I think everybody thought they were going to die, and here we were, and we were in one piece. And I kept thinking, "Darn, something must be wrong with me. I must be hurt somewhere." But, you know . . . it was amazing how uninjured everybody was at the back part of the aircraft.

MALE, SEAT UNKNOWN

The next thing I recall is a man on top of the aircraft—that is the belly of the aircraft but above me, and he was able to touch my arm. And he said, "We are going to get you out quite soon." He was very reassuring. And he called for equipment. And then it was a long period of time. And he said, "We are going to move the aircraft now and try and get you all out." Well, whatever they did at that point levered a piece of the fuselage down into my chest that had not previously been touching me. And I was really very uncomfortable at that point. And [I] shouted at [the rescuers], "You are crushing me." And [the rescuer] could see that, so they stopped. But from that point on it was very unpleasant with a lot of pressure on the right side of my chest. And I could not breathe well but rather had to splint my chest and sort of grunt to breathe and couldn't talk very well at all.

Then the next thing, the medical people were on top of the aircraft trying to start intravenous fluids on those of us [whom] they could reach. And there was a very feminine voice, I believe, who was trying to start my IV in my right arm that was up behind me, and every time she would get [the IV] started I was shaking so violently it would pull out. And this all-feminine voice after several tries said, "Every time I get it started the son of a bitch pulls it out." And if I could have laughed, I would have.

Sometime after that, they were able to get in to us from the back. And I think probably a whole tier of seats was pulled away [by their efforts]. Momentarily there was a lot of seat-belt pressure. And I shouted, "Cut the seat belt." Which they did. That relieved [me], and I was able to move away from the fuselage, and then it was marvelous.

FLIGHT ATTENDANT KELLY ENGLEHART

So, anyway, we all got into a huddle, and eventually a bus pulled up and so everybody ran to get on the bus.

You might want to remember that it was a blizzard outside, and it was very cold. So, I noticed the six-year-old little girl that we let out first. A fireman put her right away in the fire truck. So, then, after everybody was on the bus I went back to check on [the other flight attendant], and I ran into a fireman. And I said, "What can I do?" He said, "Well, why don't you run around the aircraft and see if you can see anything or help anybody."

So I ran around the aircraft, and I found an elderly gentleman lying on his back, and he was . . . like, half in and half out of the airplane. He was screaming, and he was moaning that he was freezing. So I took my coat off and put it over his hands, and I rubbed his hands and rubbed his cheeks and said, "We made it. We are alive!" And he just kind of smiled. And as I kept talking to him, I stood up and I looked over, and [I saw that] the whole wall of the aircraft on that side was gone. And there were at least ten or twelve people still in their seats upside down, and I could tell by looking at them that they were not alive.

So I just kept on going, running around the aircraft, trying to find [the flight attendant who died] to see if I could just see her anywhere. And I went all the way around the aircraft and I could not see her. Then I went back to [the flight attendant who got out with her]. . . . [He had] injured his back. And at that time the firemen had brought a board for [him] to lie on out on the snow, and he was strapped in. And he kept saying, "Don't leave me." And I said, "I have got to go be with the passengers on the bus." I thought I should be there. . . . He was lying in the snow on a board by the fire truck. I jumped to the fire truck and I got the [coat that belonged to the] guy who was sitting in there, and I threw it over to [the flight attendant on the ground]. And then I took the firemen's gloves and I put them on. And I went back down and [the flight attendant with the injured back] told me to unbuckle him, because he did not want to stay lying on the board, because he was afraid somebody was going to run over him or that he was going to freeze to death. So I started to unbuckle him, and then two firemen walked by and told him to stay down. So he laid back down. And I told him . . . because of his back injury that he should stay there or he was going to freeze to death or [he could] take his chances and get up and try to get in the bus. Eventually, he . . . came into the bus.

All the people [who] were on the bus . . . had minor injuries. At one point, they put two people on stretchers, and they slid them into the floor of the bus. . . . One of them was conscious. The other one was incoherent, was crying for help. And it was really hard for me . . . to understand why they were in our bus because we were all—nobody was hurt. And when we would look out the windows all we could see were . . . ambulances. All kinds of them [were] out there. . . . At one point a fireman came on board the bus and he was real excited, and he handed me all these tags, and he said, "Start tagging the people, start numbering." . . . I looked down at the tags and was trying to figure out what they were for, and I thought, "OK, I will start tagging." And I borrowed the bus driver's pen, and I looked down and I really . . . did not understand how I was supposed to tag or who I was supposed to tag . . . but by the time I looked up to ask [the fireman], he was gone. . . . So I just tossed them in the corner.

Two people brought [an off-duty flight attendant] over to the bus. He was completely incoherent. . . . They sat him on the bus, and I walked over to him. His shirt was ripped open. He was shaking and his face was purple. He looked like he was dying. I could not believe he was on the bus, because he looked like he, you know, [was] seriously injured. . . . I knew I took a coat and I covered him up, and I think it was about fifteen or twenty minutes later, somebody came and took him and put him in an ambulance.

8. FLUSHING, NEW YORK

March 22, 1992

USAir Flight 405

USAir Flight 405 crashed on takeoff from runway 13 at La Guardia Airport. After liftoff the aircraft's wing contacted the ground, and the aircraft veered left off the runway. The main landing gear struck a guidance light stanchion, and then touched down again. The aircraft became airborne a second time, and then hit a runway antenna and a utility building before tumbling over a seawall. The aircraft came to rest upside down in Flushing Bay, with a part of the fuselage and the entire cockpit submerged in water. There were forty-seven passengers and four crew members on board. The captain, one of two flight attendants, and twenty-five passengers died of their injuries. Instrument meteorological conditions existed at the time of the incident: indefinite ceiling, sky obscured, vertical visibility seven hundred feet, visibility three-quarters of a mile, light snow and fog, temperature thirty-one degrees Fahrenheit, winds zero six zero at thirteen knots, drifting and wet snow on runway.

COCKPIT VOICE RECORDER

From the start of the recorder at 2104:42, four minutes pass before the captain of flight 405 makes an announcement of their progress to the passengers.

CAPT: Folks, we are in line for takeoff and I see about, about seven airplanes ahead of us, so it's not going to be about . . . about another eight or nine minutes before it's our turn to go. So thank you for . . .

While waiting, inching along in the line of aircraft waiting to take off, the cockpit crew talk about deicing, in a roundabout way.

FO: You ever see that car wash they have at Denver? They like mount it on to the hardstands. That's the ideal way of doin' it, man.

CAPT: Yup.

FO: They ought to have something like that . . . in New York, you know? They ought to have [one like] that out there.

CAPT: Yup.

FO: Zip, zip, zip, man. Just you know, put it on the tab. Just cruise on out and take off.

CAPT: That's really the only surefire safe way to do it.

FO: Yeah.

CAPT: Have it an airport function. They charge each airline as they come through.

FO: Man, we pull up behind this [MD] Eighty. He might keep our wings clear [of snow and ice] for us.

CAPT: Well . . .

Sound of laughter.

CAPT: It can cause us to refreeze, too.

FO: Yeah, it's true.

CAPT: I don't want to get very close to him. How'd you like to be stopping an L-1011 out there tonight?

For the next thirteen minutes the crew talks about schedules and regulations.

FO: Look at all that stuff.

CAPT: What is that, sand?

FO: Sand, I guess.

Sounds of laughter.

FO: Aviation is my life.

Sound of yawn.

FLIGHT ATTENDANT DEBRA ANDREWS TAYLOR

There was a lady [passenger] who was concerned about making sure that the wings were deiced. She came up. I was already seated in my seat. We had already done the [emergency procedures demo]. Ready to go back and everything. And she came up, and I asked her what was wrong. And she said, "There's a lady back there whose daughter told her to make sure that we were deiced properly," basically is what she said. And I said, "OK."

So Janice [another flight attendant] knocked on the cockpit door and used her key and went in there and closed the [cockpit] door behind her. [I] couldn't hear any conversation between [her] and the cockpit crew. She came out and I just looked at her, and she says, "Everything's fine." And so I said, "OK." So she went back and sat in her seat. We pushed back. I guess . . . [we] got into position to take off.

The cabin was dark, you know—low lights. And we were there long enough to where [the other flight attendant] went ahead and made an announcement to the cabin that in preparation for takeoff please make sure that all seat backs and tray tables are in the upright and locked position. Well, when she made the announcement I got up and checked first class and she checked the back, just making sure . . . that seat belts were tightened [and] everything was up. [I] met her [at] about [the] window exits. And I turned to her, and we looked at each other—like gave each other the "OK." And I turned around and got seated, and she went back to her jump seat and belted and everything.

MALE, AGE FORTY, SEAT 5F

After boarding I changed my seat and sat in 5F. I did not bring any carry-on baggage on the aircraft, nor did anyone stow any carry-on baggage under the seat directly in front of me. The windows were very icy. The deicing was very good the first time [it was tried]. The pilot stated over the PA system, "For all you flap watchers, I'm going to leave the flaps up for safety reasons because of excess snow."

The first deicing took about ten minutes. We sat for a while. We waited twenty to twenty-five minutes at the gate from the end of the first deicing until the second deicing. After about fifteen minutes the pilot announced, "We have a strange situation. The deicing truck is stalled and we cannot back up [out of the gate]."

Before the second deicing the deicing truck came around the right wing of the aircraft. The deicing fluid ran down the windows. The deicing truck then went to the left side of the aircraft and deiced. We were then pushed back. The wings appeared to be clean after the second deicing.

I had my seat belt secured tightly around me. I cannot recall whether or not a passenger safety briefing was made. I have no recollection of flight attendants doing anything because I was busy in conversation with my friends. I was nervous. The wind and the blizzard conditions concerned me. I saw all of the aircraft lined up ahead. . . . The pilot made an announcement that there were nine or ten aircraft ahead of us and it would take another six or eight minutes before takeoff.

After thirty or forty minutes the ice built up [on the wings]. The ice completely covered the wings. It was worse than any time since I had boarded the aircraft. I bet the ice was one half inch thick. It looked very thick, but I had no real way of judging.

FEMALE, AGE THIRTY, SEAT 5E

We were all frightened that something would happen. We all said our good-byes to one another. My friend in 6F told me that she had a religious medal of St. Christopher to protect her. I reached into my carry-on to get my religious medal before takeoff. My husband and I held hands with my religious medal between us, and kissed good-bye.

FEMALE, AGE THIRTY-FOUR

I remember saying to myself [when the pilot announced that they would be taking off in eight or ten minutes or so], "Ten minutes! If we're number eight [in line for takeoff] it'll be at least twenty or thirty minutes." As we sat out there, waiting to take off, I was more nervous than I had ever been in a plane.

COCKPIT VOICE RECORDER

The crew further discusses flight schedules at 2126:39.

CAPT: We go to Greensboro and we get in at eleven fifty-three, so it ain't gonna get us home any earlier.

Nearly six minutes pass, waiting on taxiway for takeoff, and the crew goes through the pre-takeoff checks. At 2133:50 the tower contacts the flight, advising them to "taxi into position and hold on one three."

FO: Position and hold One Three, USAir 405. [On PA] Ladies and gentlemen, from the flight deck, we're now number one for departure and we would like our flight attendants to please be seated. Thank you. [To captain] Flight attendants notified, transponder and flight director's on, before-takeoff check's completed. Okay, ignition's on. Flaps eighteen. [There is] a little discrepancy in our heading of about, ah, I guess that's this grid up here.

TOWER: USAir 405, runway 13, cleared for takeoff.
FO: Cleared for takeoff, USAir 405.

Sound of increasing engine noise.

CAPT: Power's stabilized. Detent set. Takeoff thrust.
FO: Takeoff thrust's set. Temps OK. Power's . . . looks good.
CAPT: Eighty knots.
FO: Eighty knots.

Sounds of nine thumps.

FO: Vee one. Vee R.

Sounds of stall-warning beep and stick shaker.

CAPT: God.

Sound of five stall-warning beeps.

FO: Come on.

Sound of first impact, stall-warning beep. After about a second, sound of second impact. End of recording.

FEMALE, AGE THIRTY-TWO

When we started to take off, [a friend] said that we shouldn't be trying to take off without deicing again. I couldn't believe that we were trying to take off under those conditions. [My friend] said, "If we take off like this, we are all dead." Either he or I said, "We are on the plane to Hell." He said, "What is this idiot doing? He's going to get us all killed."

FIRST OFFICER JOHN JOSEPH RACHUBA

When I felt this . . . single, pronounced buffet [on takeoff roll], it was very obvious to me—you could almost feel it through the airframe of the airplane—and I glanced over at the captain's face. And, obviously, there was a very intense look on the captain's face of concentration. He was right on top of the airplane. He smoothly applied, you know, aileron, elevator rudder as necessary to maintain the desired flight path of the airplane. . . . The captain was right with the airplane.

I could see the captain actively involved with the controllability problem that he had. But you, you know, that the airplane is being controlled. I mean, you could feel it, you can see it. You can see your own control column moving in tandem with the captain's. It's very obvious. The captain had commanded the throttles that evening. He was the flying pilot. I don't recall any specific adjustments that I made to the throttles. . . . He regained control in the sense that we had an aircraft . . . that [he was] flying it out there, you know. [He] was doing everything he could do.

We had checked the wings. We had been deiced. We had noticed that the snow had tailed off to next to nothing and checked the wings.

FLIGHT ATTENDANT DEBRA ANDREWS TAYLOR

Then I just remember, basically, just sitting down in my jump seat in the brace position, like you normally [do at] takeoff. . . . I remember rolling down the runway. . . . I don't remember ever getting off the ground. I, of course—now I know we got up in the air.

But at the time . . . when I think what happened, I don't ever remember leaving the ground. I don't have windows by my seat. . . . All I have is a galley area in front of me and two doors. One of them is [the door to the entrance] stairs and one is a little catering door that's probably about maybe four feet high. . . . There's no way I can see what I'm going through, you know. I was just judging—I just kept—remember thinking to myself, we were going down the runway. It got real bumpy. Just jolting and everything.

I just remember thinking to myself, "Fellows, put the brakes on, slam ['em] on. We just don't go, just don't go."

MALE, AGE FORTY, SEAT 5F

As the aircraft started down the runway at full throttle the takeoff felt mushy. I remembered a loss of altitude. I felt a bump, a shudder, and the plane veered drastically to the left and the aircraft touched the ground.

FEMALE, AGE THIRTY, SEAT 5E

Then the pilot said that we have just been cleared for takeoff. I was not aware of any drag or shakiness of the aircraft during the takeoff roll. The aircraft went up—not too high. . . . The plane veered to the left. There were banging noises. The two loud "bams" were associated with the movement of the aircraft. Then a huge tidal wave came over us, and my husband's and my hands were pulled apart.

FEMALE, AGE THIRTY-FOUR

As we began to move down the runway it just seemed that the airplane was not moving fast enough. As the plane started to lift it didn't have that sitting feeling like all other jet flights I have ever flown on. It felt like the liftoff of a prop. We got about forty or fifty feet off the ground and the plane began to roll to the left. The pilot tried to bring the plane back to the right and abort the flight. The plane slammed to the ground about five percent off line to the runway. The plane bounced up and came down again.

FEMALE, AGE THIRTY-TWO

The plane tilted sharply to the left and right, back and forth, a few times. I thought that we would pull out of it because we hadn't touched the ground yet. The plane started bouncing up and down. Then I saw [my friend and his wife] kiss and she said, "I love you," and [he] said, "We are going to die."

Then I thought this was really serious and we really are going to crash.

FIRST OFFICER JOHN JOSEPH RACHUBA

The takeoff roll was very normal. The captain smoothly brought the throttles up into the [takeoff] position. And I said, "Takeoff thrust!" I advanced the throttles to one hundred percent on takeoff thrust indicators and stated [that] the power check's okay. [I] checked the engine instruments. The airplane accelerated normally down the centerline of the runway. And . . . the captain's rotation was normal. . . . The airplane continued to accelerate and the airplane made a notably smooth transition from takeoff roll to flight. And just as the wings leveled, just as we felt the weight of the aircraft be accepted by the wings . . . we had this encounter with this very single, long, pronounced buffet and this dramatic roll to the left and the cessation of climb.

I glanced over at the captain and he had a look of surprise and, more importantly, this very intense look of concentration. Stayed right with the airplane. And he deflected the aileron rudders and elevators as necessary to maintain desired flight path. We were in a nose-high attitude there. Somewhere along here I kind of remember the sensation of the airframe, one of the wings contacting the runway. You get a sort of a sensation, a feel through the seat of your pants. I joined the captain at some point in time on the controls. [I] placed my hands on the control column, my feet on the rudder pedals, in an attempt to monitor exactly what the captain was doing. And I . . . was impressed [that] the captain was doing everything right.

I looked outside the aircraft to observe the runway environment and the terrain and any obstructions. The airplane was . . .[starting] to leave the centerline because of the roll and the . . . captain applied the necessary control inputs and we established a pitch attitude that resulted in a loss of altitude there.

There were several attempts made to get the aircraft to climb. We had very little vertical control of the airplane. You know, when you're flying an airplane, if it's going to receive your attempts to get the airplane to climb favorably or not. I felt that the airplane wasn't going to accept any amount of excessive back pressure [on the control yoke]. I felt that the airplane would have exhibited the same tendency it did. . . . It would have stalled, broke off, would have inverted and nosed in.

Just about the point in time where you realize that you're running out of energy and altitude, you know you've got to abandon hope of flying the airplane out of there and start thinking in terms of making the crash survivable. . . . I realized and I'm sure that [the captain] realized that we're coming down, but we still needed to strive and induce an attitude in this airplane that when we did impact it would be conducive to survival.

. . . We had very minimal control over the vertical flight path of this airplane, but we did have some lateral control. Again, we were maneuvering the airplane to avoid obstructions during the crash sequence. I recall flying . . . into the water. And I recalled seeing the water and the jagged rocks and the top of this earthen dam that is very narrow and realizing that this is certainly no place to initially impact the earth.

I know the captain was resigned to the fact that the airplane was not going to fly, and crash, because he said, "God." And, you know, I believe that's when he knew that we were coming down, although he was flying the airplane. I pushed in right rudder and I felt the captain had right rudder also and I knew that even though he probably didn't see what I saw, he was following my lead. He had no choice but to do it. He saw nowhere to go. There was nowhere to set the airplane down out there that night.

. . . I was somewhat pleased with the impact. I expected it to be more severe.

FLIGHT ATTENDANT DEBRA ANDREWS TAYLOR

And I remember a big jolt. I mean, as well as I was strapped in, I remember a jolt to where I went forward in my seat and the galley area to where there's in front of me probably about three feet or four feet in front of me—it's a big area with a lot of the carts there and a lot of heavy equipment. I remember being jolted. I remember someone screaming. And then just getting tossed around and torn around and finally kind of stopping. And then when it kind of stopped, I remember a gush of water coming in on me. At the time when I was sitting there . . . something had encompassed me, had come over me.

Nothing was touching me. Nothing hit me in the head. But I could tell there was something around me, wreckage around me, or something around me. And we were in the water. And I remember being submerged in water and thinking I was going to die. I sat there, and I really thought that this was going to be it. I just closed my eyes, thinking I was going to see that bright light at the tunnel and gave it a chance. And it never came. So I thought, "Well, I'm getting out of here." So I just kind of got up.

MALE, AGE FORTY, SEAT 5F

I saw a flash. I saw a fireball to the left and front of my seat after the crash.

I was still in my seat on the right side of the aircraft. I heard the aircraft break up. When it stopped, a muddy slush and water came over me. The water was up to my nose. At first I did not think about my seat belt. I had difficulty removing my seat belt. The seat-belt buckle was behind me and to my right. My wife, who was seated next to me in 5E, encouraged me to keep trying. I had swallowed a lot of jet fuel and was feeling ill and I wanted to give up. After several attempts I was able to release my seat belt.

FEMALE, AGE THIRTY-FOUR

I saw a bright red-orange flash out my window. The plane then went into a roll or flip. It's hard to say because it was so fast and violent that all relationship to up or down was hard to determine. When the plane finally settled I was upside down, under water, and had the feeling of being trapped. There seemed to be things across my back and things to the side of me that had me pinned with my face against what I think was the window wall. I struggled to free myself, but my efforts seemed thwarted.

FEMALE, AGE THIRTY-TWO

The plane started skidding and exploded. People started screaming. I heard a scraping noise as the plane was skidding on its belly. I saw a yellow light and felt a tidal wave of heat and wind, which was the explosion from the back of the airplane. I closed my eyes and bent over. We were flying all over the place. I hit something while we were in motion. I came loose [from] my seat and my body was bouncing around like it was in a [clothes] dryer. Something hot went up my back. Something pulled my sweater over my head.

Then I found myself down on my knees in very cold water. Everything was black.

FIRST OFFICER JOHN JOSEPH RACHUBA

• And just about this time I recall the fireball. I knew that [the fuel cells] had exploded.

And the airplane slid along the ground . . . what you might imagine a belly landing to feel like [with] the surface of the earth passing underneath this skin of the airplane. Then . . . I remember an extremely violent impact with something outside the aircraft. And, as a result, I was thrown to the right and remember some pain and discomfort as the airplane came into contact with this obstacle.

And then the brief sensation of weightlessness as the airplane was launched up over the top of this obstacle. I learned that it was probably a dam out there, a dike. And then a view of the shoreline again, the water and the rocks and just some, you know, puzzling on my part as to why am I seeing this shoreline again? I thought we had flown over that. But I assume that's when the airplane went up over the dike and then inverted down into the water.

. . . And then at some point in time I remember this feeling, this knowledge, of just total disorientation. This whole ordeal for me personally is working down into two units: the crash and being in the water. And I had this recollection of just total disorientation. I didn't know where I was. I didn't even know I was aboard [the aircraft]. It was pitch-black. And I thought that at first, you know, I was home having a nightmare. It was very much like that.

And then after I started to realize that everything I was seeing and hearing and smelling and feeling was too real to be a dream, [and I] thought I'd [awakened and] gone to hell or something, in my sleep. I then started to realize, and the stark realization . . . set into me, that we had, in fact, crashed.

FLIGHT ATTENDANT DEBRA ANDREWS TAYLOR

I remember [being] frustrated. I couldn't get up. I was scared. I was getting scared. And then . . . I thought, "Oh, God, my seat belt." So I reached down and got the seat belt and got out of it. And that kind of popped me up a little bit, you know, so I could breathe.

And I remember my door was right here to my right. I knew I just had to get out that door. I never opened the door. I don't know how I got out of the aircraft. I was trying to reach over here. Something beside me came floating by or just came by. Over here like, almost like a *schlump* on the side of me. It seemed like something floated by and kind of blocked me here.

From that point I don't know how I got out of the aircraft. I remember being—I don't know. I just kind of shimmied out or like a little mermaid or something, just somehow got out of the aircraft. I was real close to the aircraft at this time. And I was aware that I had to get away from the airplane. I knew it was going to catch on fire. . . .

MALE, AGE FORTY, SEAT 5F

I recall seeing small fires, bent metal, and a hole in the aircraft. My wife was too frightened to jump out of the aircraft. My wife and I went out of the hole [and] into the water.

FEMALE, AGE THIRTY, SEAT 5E

I found myself free of my seat belt without removing it myself. Something hard was on my head. I pushed with my feet to get up before the water rose. I helped my husband to get up. I tried several times to unfasten his seat belt but I wasn't successful. He was able to release his seat belt himself.

I told my husband to get out.

FIRST OFFICER JOHN JOSEPH RACHUBA

. . . The immediate thing I remember was just the presence of water all around. It was up to my chin, and getting in my mouth and down into my lungs and into my stomach. And as a result I was coughing and hacking on it. And I remember the odor of jet fuel and I think probably one of the things I remember most was just the filthiness of the water. It was like walking into a farm in the summertime and stirring up this black muck and it seemed like more muck and filth than water, actually.

But I remember [I heard] some [calls] behind me. People [were] calling for their companions. I remember this moaning and groaning, this very low moaning and groaning. I thought—I didn't know who it was. I thought it might have been the captain. I learned later it very well could have been me moaning and groaning and not realizing it.

. . . Water was flooding the cockpit and there was just a very small area between . . . the water, which was up to my chin, and the surface of the airplane. And my first fear personally was that of drowning. And I thought the airplane was going to sink. I could see through my eyebrow window. I could see the moonlight reflecting on the surface of the water outside the airplane. And I didn't realize until just a few months after that the airplane had actually come to rest on the riverbed, on the surface—on the bottom.

I surveyed the cockpit and it just . . . it was not representative of the F-28 cockpit the way I'm used to seeing it. It was dark in there. There was water everywhere. It was very cold.

My first conscious effort was to try to locate the captain. I called out to him, "Wally, Wally." I didn't hear any response back. As I started to make out some things in the cockpit I could see the captain's leg suspended above me, not draped, but touching, but the way we came to rest he was still in the seat strapped in, and I could see his leg and I tried to follow his leg to where the rest of him, I thought, would be located. And he just seemed to disappear in a mess of twisted metal. And I could tell by the way that he was reclining that, unfortunately, the captain's head was below the surface of the water. Then I concluded that he had died.

Knowing that there wasn't anything I could do for him at that point, I needed to get out of my seat. . . . I was in water up to my neck. . . . I reached down with my right hand and attempted to open my seat belt and had great difficulty because my right hand was frozen and I couldn't make a fist. I banged on it about eight or nine times and I got out of my seat, where I dropped down into my sidewall area and started thinking that I needed to find a way out of here.

[I] started looking around and saw no way out. I reached for the sliding window, my sliding window, and it's like Murphy's Law. I went to pull on it and it didn't open. I tried several times and even if it had opened it would have been no way to egress the airplane because the airplane was actually resting on [the window]. I turned around and looked for another way out. I couldn't find a way out.

. . . And I realized that, you know, I'm alive, but maybe for a few moments, when I got to the end of my rope, that's when I pretty much got serious with my own personal situation and I really had nowhere to turn. And I called on my Lord for help, and I said, "Jesus, help me, Jesus, please help me out of here." And this was to me very interesting. I felt this very strange but familiar presence right there in the cockpit with me. And I just felt this fear going, [and] it was replaced with peace.

And then I didn't remember anything for a while. I don't remember how I got out of the airplane or anything. [But] right about that time, I saw some rescuers outside.

FEMALE, AGE THIRTY-TWO

I went left into an upside-down metal tepee-shaped area. I couldn't see a way out of there and thought, "Don't stop moving. I've got to find a way out!" I saw a light behind me from the inside of the plane and I went in there. The light was from the fire. I smelled smoke and heard a hissing sound. The smoke was so thick that I was choking and felt like I was going to pass out. There was a woman lying in the water. I helped her get up. Pieces of the fire fell down on my hair, singeing it. I was screaming for my husband. I could hear people moaning. I saw the stars and knew I was outside. There was an older man standing there. The three of us walked through the middle of the plane and out between a piece of fuselage and tail. The tail was on my left. There was a guy in the water screaming, "Help me!" He was all bloody. I became worried about getting AIDS. I bent down to help him but he said, "No, the plane is on my back." I felt that I couldn't get him loose by myself. Another man and woman started helping. A fire started next to us and I put it out. Another fire started and I put it out also. With all the fuel in the water, I was afraid that the water would catch fire. I yelled and looked for [my husband].

FIRST OFFICER JOHN JOSEPH RACHUBA

. . . I went out this hole, headfirst, into the water, under water, and managed and tried to start swimming. And there was some guys out there to greet me. And they ran over, grabbed me by the arms and walked me up around the nose . . . of the airplane, up onto the shoreline.

FEMALE, AGE THIRTY-FOUR

Finally, I was able to get my seat belt released and was able to get whatever was on me off. I came to the surface [of the water] in what I think was the luggage compartment. There were wires and hydraulic hoses and a mass of ripped and torn metal. There was a hole in the side of the wall of the plane that I was able to get through. I was on the bay side of the plane. I had swallowed water and jet fuel, so it was difficult to breathe. There was fire at the rear of the plane and suddenly it flared up. I went back into the water [that was] chest-high. I waded to the [shore].

FLIGHT ATTENDANT DEBRA ANDREWS TAYLOR

I remember kind of—my left leg was broken. And of course at that time I knew my leg hurt and it was cold. And I lost my shoes and I was wet and everything. But I was on a rock and I knew I had to make my way closer back. So I was kind of crawling along to another rock which was further away from the aircraft. And I sat there on the rock. And I was cradling my leg because I knew my leg was very very cold. And I didn't know whether it was going to get cut off or not because it was swollen and it was blue and it was scaring me with the circulation and everything.

So I was holding my leg in my arms, kind of cradling my leg. And just sitting there, watching the plane on fire, watching the plane blow up and it was on fire. I remember seeing people. From where I was there was no one around me as far as passengers. I was away from the aircraft.

There was a lady behind me on some rocks. And she was walking behind me, and she asked me if I had seen her husband. Just normal. She said, "Have you seen my husband?" I said, "No." She said, "Oh. I'm going to miss the family reunion," or something like that. . . . I said, "I'm sorry, I haven't seen your husband."

And I'm watching people jumping out of the windows, window exits. . . . They were just running everywhere. It wasn't like one at a time. They were just kind of just dispersing everywhere. And like I say, I was there holding my leg.

I remember seeing, at one time—at one time I looked at my watch, when I first got on the rocks. And it was about ten 'til ten. I did look at my watch. And I remember I was mad because I didn't see any help. I didn't know if they even knew about us or not. I didn't see any rescue vehicles. So I was just getting mad about it.

FEMALE, AGE THIRTY-TWO

At this point I was on an island-type thing. Another passenger who was standing on a portion of the plane yelled, "Is the water deep?" I told him that it wasn't and he came over to where I was. A flight attendant was yelling that her leg was broken. . . . I was scared to go back in the water. A guy from the land, who turned out to be a snowplow driver, was yelling for us to come over to him. He said to get away from the plane because it might explode. He came over to us in the water and gave me his coat. I got into the water but my feet were numb, so he helped support me as I went through the water. He helped us up this big cliff. I was screaming [for my husband]. [My friends from the airplane] came up and asked where [my husband] was.

It was so cold. I had no shirt or shoes on, but did have on the coat that the snowplow driver had given me. The wind was blowing and it was snowing. We sat in a circle hugging each other, wondering how long will this take? My feet were so cold that I couldn't stand up anymore. I sat down on my feet to try to keep them warm. We thought that we were going to die of exposure. It seemed like a half an hour until some firefighters showed up. They were telling everyone to get away from the airplane. I told one of them that I couldn't walk because my feet were so cold. I slid down the hill on my butt for a while. The firefighter left. I became separated from [my friends]. I looked for [my husband], checking the other people. I saw [my friends] again by the fire truck. We decided to stay together so that [my husband] could find us.

MALE, AGE FORTY, SEAT 5F

My wife prodded me to go and climb up the seawall and over the embankment because I was feeling sick and did not want to climb the wall. I was assisted by another crash survivor at the embankment to get on the shore. I walked to the utility truck and we stayed there.

FLIGHT ATTENDANT DEBRA ANDREWS TAYLOR

And then I remember one time seeing John [Rachuba, the first officer] walk by in front of me. I knew it was John because of the epaulets on his shirt. And he just looked like he was walking kind of aimlessly in front of me. And I called to him. He didn't answer me or anything. And I didn't have the energy really to yell. I didn't really want to say anything. [I] just wanted to be quiet.

So I sat there for a while and then another fellow, his name was Yasser, came by and asked me if everything was OK. He was a passenger. And I said my leg kind of hurts. So I lifted my leg up and he looked at the leg and it was just kind of going down [the swelling]. There wasn't any structure to it. So he lifted me up, and I held his neck, and he just carried me over some hills . . . and up an embankment. It looked like a truck, a catering truck, an empty catering truck, and [he] put me in the back of a catering truck in the far-left hand corner, I assume to shield me from the weather.

And he wanted me to stay there and not move and he said [he would] get some help. So he went and got some paramedics who had a stretcher. They came by with the stretcher and put me on it and carried me [to] an ambulance.

FEMALE, AGE FORTY-NINE

In a situation such as this, I don't think much matters. It's amazing any of us survived.

9. *LOS ANGELES, CALIFORNIA*

February 1, 1991
SkyWest Flight 5569 and USAir Flight 1493

SkyWest flight 5569, a Fairchild SA-22, was to depart runway 24L, Los Angeles International Airport, in the early evening (6:04 P.M.). Taxiing from the gate, SkyWest's flight crew requested departure from midfield taxiway 45, and the local controller cleared the flight onto the runway, requesting the flight to hold for crossing traffic downfield of where flight 5569 was holding. The crossing traffic on runway 24L was Wings West (WW) flight 5006. Interrupted communications between WW flight 5006 and the local controller caused a delay in crossing runway 24L and, therefore, a delay giving permission to SkyWest flight 5569 to take off.

Meanwhile, flight 1493, a USAir Boeing 737, on final approach, was cleared to land on runway 24L. USAir flight 1493 collided with SkyWest flight 5569, which was still holding, waiting for permission to take off on the centerline of runway 24L at the intersection of taxiway 45. The collision of USAir 1493 with SkyWest 5569 created an explosion and a fire. The two colliding airplanes slewed off the runway and slid into an unoccupied fire station. USAir flight 1493 evacuated sixty-five passengers, three flight attendants, and the first officer. Twenty passengers and two crew members of USAir flight 1493 died. All ten passengers and two crew members on SkyWest flight 5569 were fatally injured. As contributing factors to the accident, the National Weather Service reported that official sunset and the end of official twilight for the Los Angeles area occurred at 5:23 P.M. and 5:48 P.M. respectively.

Before the evacuation of USAir flight 1493 the cabin filled with "thick black smoke." A female passenger in seat 10F "froze" and was subsequently unable to get out of her seat or open the window exit next to her. A male passenger in seat 11D climbed over the 10E seat back and opened the over-wing exit and pushed the female passenger out of the window onto the wing and then exited himself. During the subsequent evacuation through the right over-wing exit two male passengers had an altercation resulting in a fistfight that lasted for several seconds. Meanwhile, at the over-wing exit door passengers pushed and kicked and shoved and "fought" one another in a panic to get out.

COCKPIT VOICE RECORDER

The cockpit crew of USAir flight 1493 engaged in relaxed, amiable conversation prior to approach to LAX.

FO [to passengers on PA system to cabin]: A note from the flight deck. We're on our final descent into Los Angeles at this time. About another fifteen minutes for the approach and the landing. Current temperature is sixty degrees under a little bit of cirrus clouds. Our thanks for joining us today. I'd like to take this opportunity to wish everyone a very pleasant weekend. At this time I'd like to ask the flight attendants if you would prepare for our arrival. Good day.
FO [to CAPT]: About . . . ah, that rate of descent is set up for eleven thousand.
CAPT: That's right?
FO: Ah, yeah, it is.

Nearly four minutes pass.

FO: Cleared for visual for 24L.
CAPT: Left?
FO: Correct.
CAPT: I'll confirm that. [To LAX Approach Control] Ah, just confirm the visual for USA 1493 is to two four left.
LAX APR: That's correct, USA 1493.

The crew configures the aircraft for landing.

LAX TOWER: USA 1493, cleared to land runway two four left.

CAPT [to LAX TOWER]: Cleared to land two four left, 1493.
FO: Looks real good.
CAPT: Ahhh, you're coming outta five hundred feet, bug plus twelve, sink is seven. [Ten seconds later] Lights on.

Sound of impact. End of recording.

FIRST OFFICER, USAIR FLIGHT 1493

The flight was very long. Topic of conversation was naturally the war going on [in the Persian Gulf]. I think that was on a lot of people's minds. I know it was on mine, also being a member of the Air National Guard. We talked quite a bit about our children [and] Christmas vacation down in Florida and just small talk about my kids . . . things with the war. That was pretty much it during the route.

Somewhere during the route, we also got served a meal. [One of the flight attendants] had informed us back in Columbus [Ohio, where the flight originated] that there were some extra first-class meals, so we were excited about that, I guess you might say. Four-and-a-half-hour flight, it's always nice to have a little to eat. Like I said, the leg went very smooth. No problems. The airplane, it was fine. No problems with the airplane.

Somewhere, I would say approximately one hour from Los Angeles, the sun had started to go down. I was able to remove the sun visor and also take off my sunglasses at this time, although it was still definitely daylight out but [I] really wasn't having too much problems with sun glare.

At approximately twelve miles out we were on glide slope at about 250 knots and slowing, and at this time I went ahead and asked for gear down [and] before-landing checklist. . . . I'm in the process now of slowing the airplane down, clearing as much as possible. Very much aware that there's four runways and potential for a lot of air traffic. . . . I'm very conscious of [our] distance away from the runway . . . what the winds are doing to me, and just exactly how far out I am from the end of the runway. I'd say at about twelve miles or so is when I really started to get a real good feel for [runway 24L], picking up the runway edge lights, centerline lighting started coming into view. . . . Taxiway lights . . . are starting to come into view also. Everything's progressing normally.

We contacted tower for landing instructions 24L; no answer. At this time we were in final landing preparation. We were working on the before landing checklist and notifying the flight attendants [with three bells] of landing. Once again, we requested landing instructions, but no response. Tower was talking to other aircraft on the ground. I remember ATC asking the position of aircraft on the ground but cannot remember the exact call signs [of those aircraft]. At approximately three to four miles [out] we asked for landing clearance for 24L. Finally, landing clearance was given to land 24L. Aircraft was fully configured [for landing]. . . .

In our cockpit our duties are starting to pick up. We're starting to get more active. More involved with the primary duties of flying the airplane at that time, concentrating and navigating ourselves to the proper runway and good alignment and whatnot. Somewhere during this, I went ahead and asked for, I believe I said, "Flaps one," and went ahead and selected flaps one, then flaps five. . . . Now, the airplane is slowing down at this time. I'm conscious of air speed restrictions that I have for my flap limitations and things of that nature and basically just run the checklist and clearing also.

. . . I was starting in the process of the point of thrust reversers and come in gentle on the brakes, [and] that's when I saw in my one screen the silhouettes of two props quickly, rapidly fill in the windscreen. It was a split second after the nose wheel had touched the ground that the airplane all of a sudden showed up out of nowhere. I also remember seeing a red light on the tail. It seemed that, as far as position in the windscreen, it just seemed maybe I was slightly looking down at it. Immediately upon impact the sound was almost overwhelming of two airplanes colliding, of crunching metal. There was an explosion.

FLIGHT ATTENDANT PATRICIA HODGES, USAIR FLIGHT 1493

. . . All of a sudden, we just landed, and the first thing that went through my mind is [that] it is a good landing, you know. It was a decent landing. It wasn't no jumping up or down or anything.

And then all of a sudden, he, like, slams on the brakes, and it was like if you were test driving a car, and you slam it on, and then all of a sudden we hit, and then we heard this big loud noise when we hit, and then we looked out our window, and fire was coming out both sides, and we saw it in our visual thing, and the back galley was orange.

FIRST OFFICER, USAIR FLIGHT 1493

The airplane seemed like it nosed down. It almost seemed like we jerked forward from the impact. Immediately trying to analyze exactly what had happened, at the same time we departed the runway surface very rapidly. It was like it was dark. I don't remember seeing any engine instruments. No indications in the cockpit. As we departed the runway, heading toward the left, rapidly I could see the windscreen from left to right fill with a building. I could see [the captain's] hands attempt to shut down the engines. I could see a downward motion of his hands trying to get to the Start levers to cut them off. I'm doing everything I can to get that airplane stopped but it won't stop. I could see the building coming from the left and I know I want that airplane to go to the right. It wouldn't go to the right. And it was very quick. It wasn't a long, drawn-out time. It was a matter of seconds after we hit that airplane that we hit the building, and we impacted the building very, very violently, and it just seemed like it stopped. It didn't keep on sliding or anything. It just, the plane just stopped.

FLIGHT ATTENDANT WILLIAM IBARRA, USAIR FLIGHT 1493

I remember the landing was very normal and smooth. That's about all the time I had to think before the first impact, where I realized that something was not going right.

On the first impact I heard a big metal scrape and the cabin lights went out. The emergency lights came on. I remember the cabin felt really warm, and I remember seeing smoke coming from underneath the floor right in front of the jump seat where I was sitting. I remember seeing smoke and fire from the top of the valet closet, which is located right in front of the forward jump seat, in between the first class section and where I was sitting. I remember seeing passengers not out of their seats but looking out in the middle of the aisle. A big percentage of the passengers were looking at me. As soon as the emergency lights came on I remember [one of the other flight attendants] started shouting our commands to ask the passengers to grab their ankles, heads down, stay down. We continued . . . doing this. Throughout the first impact the cabin was getting warmer and the smoke was getting a lot denser. It was a very thick smoke, and the smell of it was very hard to breathe. The plane was moving in a faster motion. We were still in our brace position asking passengers to grab their ankles, heads down, stay down, when the plane . . . came to a stop.

FEMALE, AGE TWENTY-FOUR, SEAT 4D

. . . Two or three seconds after we hit the ground the brakes were slammed on and the airplane sort of stopped some. I saw the [flight attendant] in the aisle standing up, shouting, "Stay in your seats!" I kept my seat belt on. The airplane was real bumpy and real quick, about three seconds [later] there was the final crash. The lights came on, I grabbed my cousin's hand and said, "Don't let go of me."

FLIGHT ATTENDANT VANCE SPURGEON, USAIR FLIGHT 1493

. . . We then had a second impact. I fell back into the galley. I was lying completely down. I jumped up immediately, got the door pin, opened up the door—finished opening up the door. The slide inflated. And at that time my back was facing forward. I was facing aft. And I just started yelling commands: "Release seat belts. Get out. Release seat belts. Get out. . . ."

FIRST OFFICER, USAIR FLIGHT 1493

. . . The smoke was just filling the cockpit. It was very thick, burning smoke. I could see flames on the right side. I can feel the heat on the floor, on my feet, underneath me. At this point, I saw an opening over the top of my head. I thought the top of the airplane was ripped off, but there was an opening and I unbuckled myself and tried to lift myself out. . . . It was just totally dark. I don't even remember seeing the [steering] yoke that I had my hands on. I lifted my head as I attempted to get fresh air, and my leg was stuck, and it wasn't coming free, but I could get my head up to the hole [to breathe]. I could see, I could feel the clean air. I tried this a couple of times, just pulling as hard as I could. It seemed like my ankle, my left leg was trapped on something. Something was up against the left side of my leg and my ankle seemed to be twisted somehow, maybe around the rudder pedals or behind the yoke. I tried maybe three or four times to get myself out, and I wasn't able to get out, and at that point I thought I was going to die. And, like a miracle I had a vision of my wife and kids and at this time it just seemed like I had some inner strength that I can't really explain and I knew I had to get out of that airplane, and I made one more effort and my shoe came off and my leg came free, and I lifted myself up to the hole now totally exhausted, confused. I'm burning from the smoke and whatnot and trying to breathe the fresh air. I remember looking down and I could see flames underneath the airplane. Also the whole front of the airplane was completely caved in like a tin can. I could see jagged metal all underneath my window. I looked back to the right and I could see people exiting out of the galley door. I remember seeing . . . one of the flight attendants stick his head out and jump out.

· FLIGHT ATTENDANT PATRICIA HODGES, USAIR FLIGHT 1493

I got up. We thought we were just coasting, trying to come to a complete stop. And [another flight attendant] was up, and he went to his exit and I went to my exit.

And we tried to open our door, and then we slammed again, and I went through the cabin, and [the other flight attendant] fell flat down with his head going toward my door, and then we stopped again. And we came to a complete stop, and then [the other flight attendant] went to his door and then I went to my door, and then I opened my door, and before I opened my door I looked through and I didn't feel anything but I saw some flames and I didn't feel anything. So I opened my door and I looked down, and I saw flames coming down like toward where my chute was. And like it was, it was closed. And I shut it because there would be no way we could get people out on the slide. So I helped [the other flight attendant]. First, I was going to go through and try to help people at the window.

FEMALE, AGE TWENTY-FOUR, SEAT 4D

We started walking down the aisle to the back. I don't know why we started to the back. I guess I just knew there was an exit there. The airplane started filling with smoke. As we walked down the aisle the airplane filled with smoke, and it got pitch-black. I knew there was an exit. I laid on my stomach across the seats and stuck my head out of the window. I let go of [my cousin's] hand and crawled out on to the wing.

FEMALE, AGE TWENTY-SEVEN, SEAT 8F

The airplane stopped and [I] looked out the right window and saw a wall of flames. [I] released my seat belt—the seat seemed to come off tracks—[and] made my way to the over-wing [exit]. Many people were in the area. There was lots of pushing and shoving. Smoke almost instantly. There was lots of pushing and shoving. My hair was pulled. The passenger in 10F was still sitting there stunned. She wasn't opening the exit. Another man opened the right over-wing exit. I got out on the wing and was pushed off, I don't remember if on purpose or by accident. This threw me off balance and I landed forward.

FLIGHT ATTENDANT WILLIAM IBARRA, USAIR FLIGHT 1493

At that time, I remember [one of the flight attendants] and I both got out from our jump seats, kicked the jump seat. I went to my primary exit, which would have been the right-forward door, the galley service door. We were asking passengers to release their seat belts and get out. At that time, the smoke was getting denser and thicker. I remember I walked over to my door and I was assessing my door for heat, fire or water outside of the aircraft, and the way we assess the door is by going around the seal edge in a circle motion from top to bottom. I didn't feel anything that would endanger me from opening the door. I remember looking out the porthole to make sure there wasn't any fire outside, and that was the last thing I remember seeing because the smoke got so bad I couldn't see anything more after that.

By that time, I had grabbed the handle and I remember I rotated the handle to get the door open, and I started pushing the door, which was very, extremely difficult to get open. I was struggling with it, and I pushed it and pushed it, and I got it open just about ten inches where—at that time I didn't have any power left, and I couldn't breathe anymore. I was still shouting my commands, asking people to release their seat belts and to get out. I stuck my head out to breathe. The door came back on me and cut my ear, which I didn't know at the time. I then continued opening the door. Finally, I pushed it out and I got it open somehow. I remember opening the door all the way, completely open, putting the door against the fuselage, where I remember seeing the first officer climbing out of the cockpit window. My vision was very low at that time due to the smoke that had gone into my eyes, but I do remember seeing the First Officer climbing out of the cockpit window, and I remember seeing the outside of the aircraft; however, you couldn't see anything inside the cabin.

I then looked down to inflate my—to pull the red handle which would activate the inflatable chute and nothing was there. I grabbed [the] handle we have right next to the door to hold on to to make sure the passengers don't push you out.

FLIGHT ATTENDANT VANCE SPURGEON, USAIR FLIGHT 1493

. . . People were just immediately there [at the exit]. I just remember them being there immediately, and they were continually jumping out. I only remember one man hesitating, and it wasn't very long. And we just kept yelling our commands and shooing them out of the plane.

FLIGHT ATTENDANT WILLIAM IBARRA, USAIR FLIGHT 1493

I looked back and the first passenger came up and stood by the door. I grabbed his arm and I helped him out of the aircraft knowing that there wasn't a chute. The flames and the heat inside the cabin were much more dangerous than jumping out of the aircraft. I remember seeing the second passenger jump out without any help. I continued . . . asking passengers to release their seat belts, get out, come this way, get out. No one else was around. I do remember looking back to the cabin, turning my head, and I couldn't see anything. I started walking back toward the cabin to help out more passengers. I took about three steps and I didn't make it up to row 1. The heat and flames were very intense. I felt the heat all around my body, and I turned around and I felt my life was in danger and I jumped out of the aircraft.

FLIGHT ATTENDANT PATRICIA HODGES, USAIR FLIGHT 1493

[During evacuation] I had all this energy and we were screaming, we were at times pushing people out, and I was very upset because I didn't think [the passengers] were moving fast enough. I was like, "Don't you realize what has just happened?" And we kept screaming and pushing people out, and they were very calm or in shock, either one.

So we were all screaming, trying to get people out of the door, tell them to come this way and leave everything, just . . . we kept screaming, "Come this way, come this way, come this way."

People were . . . I was trying to tell them to [leave through the emergency exit] one at a time . . . and there were like two of them trying to get out at one time, and I couldn't even get through there.

. . . I was trying to scream and tell them to come, and it was like they were just dumbfounded on that exit, that they saw it, and it—I mean, I kept saying, "Come this way, come this way, come to the back." And I was screaming for everybody to come here, and they just—nobody looked at me. I mean, briefly, I mean, I wasn't like staring at everyone for a long time. It was just like I kept telling everybody to come, and nobody else came.

So there was nobody else coming, and then I tried to get through, and then it was like so dark. I couldn't see anything. It was like on the worst foggiest day you could ever imagine, and you just couldn't see anybody. I saw the people going, the lights coming through the window exit, the people crawling out, and I was trying to scream at them, telling them to come to the back, and nobody would come. So I was trying to get through to try to see if I could help them, and then I got smoke, and I got dizzy, and all I remember is that I went to the door and I just jumped.

FIRST OFFICER, USAIR FLIGHT 1493

And just then the rescue unit showed up and they saw me. I waved to them. They pulled their trucks right up to the airplane and the first thing the guy did in the turret was open up his hose on me and about drowned me, pushed me back into the airplane. . . . I finally got back up and at that time there were three members of the rescue team at the window, climbing up the ladder, and one of them grabbed me around the waist and pulled me out of the cockpit. And I remember, once they got me out of the cockpit I'm sitting there yelling to them that the captain is still in the airplane and that there's eighty-nine people on board and there's four flight attendants. And they got me down the ladder and started walking me away and I tried to tell them just leave me, just drop me here and go back and help the people.

FLIGHT ATTENDANT PATRICIA HODGES, USAIR FLIGHT 1493

Then I ran over to the window [exit on the wing] where people were coming down, and I was trying to get them because there were these two people that were just like in shock. They were standing there [on the wing] looking, and then the man came with the fire extinguisher and I was soaked and they were soaked and I was afraid that they were going to fall off [the wing].

FEMALE, AGE TWENTY-FOUR, SEAT 4D

I jumped off of the wing toward the back of the airplane. A man was there. He hugged me and said, "Help this woman." There was a lady sitting on the ground coughing. I held her and we walked about twenty feet and then I turned around to look for [my cousin]. She was walking toward us with foam on her face. We grabbed each other and we started walking away to where everyone was congregating. I just wanted to call home.

FLIGHT ATTENDANT PATRICIA HODGES, USAIR FLIGHT 1493

. . . After that we waited for someone to pick us up.

On the bus, I did see women . . . some women had their purses, [and] that's normal, but I did see somebody have one little bag, and I was just, like, "Why did you bring your bag?" I mean, I was pissed off. I mean, I was just in a mad stage at that time. I didn't say anything to them, but it was just like, I can't believe you brought your bag.

FLIGHT ATTENDANT WILLIAM IBARRA, USAIR FLIGHT 1493

I remember getting up from the tarmac and looking at the over-wing exit where passengers were deplaning. I then remember seeing passengers standing around the aircraft and, not knowing how bad my injuries were, I continued on assisting passengers, as we're trained to do, to gather people away from the aircraft.

10. HONOLULU, HAWAII

August 7, 1997

Delta Airlines Flight 54

Delta flight 54, a Lockheed L-1011-385-1-15, had trouble leaving Honolulu International Airport, first with an "area overheat system" that was inoperative when the crew members arrived at the aircraft. The flight departed the gate at 4:27 that afternoon and taxied to runway 8R. But during the taxi out, the left wing "duct fail light" illuminated in the cockpit. The captain ordered the flight to return to the gate for maintenance, arriving at 4:55 P.M. The passengers deplaned, crews repaired the faults, and the "testing" began at 5:22 P.M. All three engines were started, and the aircraft, without passengers, was taxied out to runway 8R for an engine-run test. The aircraft was returned to the gate and at 6:45 P.M. the passengers and crew again boarded flight 54. During the initial part of the takeoff, dozens of soda cans tumbled out of bins in the left storage cabinet. The takeoff roll continued, until the aircraft reached a speed of 155 knots, when the captain noted that a "door caution light" illuminated on his dash/warning panel. The captain continued the takeoff. About a second later, he felt the aircraft vibrate, shudder, and then begin a yaw to the left, and he heard what he described as "popping sounds." [A tire had blown from excessive overheating due to the cumulative distance that the aircraft was made to taxi back and forth to the terminal before takeoff.] At this point, the captain chose to abort the takeoff, while traveling at 165 knots of ground speed and with 6,000 feet of runway remaining. He slowed the aircraft using full braking and reverse thrust. The aircraft came to a stop 164 feet short of the end of the runway. A brake/wheel fire resulted. The passengers were evacuated. Two emergency exit doors failed to open. One passenger sustained serious injuries while evacuating, 59 received minor injuries, and the remaining 245 passengers escaped without injury, as did the crew.

GROUND/TOWER CONTROL TAPE TRANSCRIPT

DELTA 54: Ground, Delta 54 is ready to taxi.

GC: Delta 54 heavy, runway 8R, hold short of runway 8L. Taxiway Romeo, Bravo.

Flight 54 taxis out over next six minutes.

TOWER: Delta 54 heavy, Honolulu Tower, good evening. Cross runway 8L, runway 8R, taxi into position and hold. Traffic will depart the parallel runway.

A further seven minutes go by, taxiing.

TOWER: Delta 54 heavy, cleared for takeoff, 8R.

Delta 54 begins takeoff roll. One minute and three seconds pass between takeoff clearance and Delta's next transmission.

DELTA 54: Tower, Delta 54 heavy's aborting [takeoff].

TOWER: Roger. Need any assistance?

DELTA 54: Tower, Delta 54 heavy's going to need assistance out here. We're hot.

TOWER: Roger, sir. I have the fire crew on their way. Delta 54 heavy, [we] have the emergency crew on their way out now, and Delta 54 heavy, uh, Tower. It appears that there are flames underneath the airplane.

MALE, AGE SIXTY-SIX, SEAT 3D

We were at the end of the runway for our departure when the captain stated that the instrument readings were not to his liking and returned us to the gate. We deplaned thirty minutes later while the plane was checked. They took the plane away from the gate and returned it a couple of hours later. We boarded and four and a half hours after our scheduled departure we taxied out to the runway. En route the captain stated that, "They have addressed the problem," and he hoped they fixed it!!! He then stated that we *should* be OK. He was obviously concerned and doubtful, and not reassuring at all. He hesitated on the end of the runway before attempting takeoff.

FEMALE, AGE THIRTY-SIX, SEAT 4F

It was extremely hot on the plane. The stewardess occasionally used an item of instructions as a fan because she was sweating.

FEMALE, AGE THIRTY-THREE, SEAT 12G

I got on the plane [the] second time . . . I was very skeptical getting on the airplane, because of something that was to take five minutes to repair (we were told the repair would take five minutes and it took three and a half hours). . . . The plane was still hot inside. As we were leaving the gate, the pilot said on the PA system, with a vague voice, hopefully everything has been checked and we'll try to make this flight a pleasant one.

FEMALE, AGE THIRTY-FOUR, SEAT 16J

We all sat out there on the runway for a long time, at least an hour, and finally they told us that something "mechanical" was wrong with the plane. I knew something was wrong, and I was glad to get off the plane. I just knew they would get us another plane, since we had to cross the ocean and fly so far to get to Atlanta. Well, we had about a three-hour layover, and then they stated it was time to board the plane. They stated we were getting on the same plane. I felt sick right at that point. I knew that the plane was not safe, and I felt like they risked us on this plane. While boarding I told the flight attendant that I did not feel the plane was safe. I was tearful by the time I sat in my seat. To make matters worse, the pilot . . . stated, "The plane is ready to go this time, I *think*." I was very angry. That was not funny. To fly that far, you should be sure, with 295 people's lives in your hands. I was angry and scared. I held my husband's hand and told him to say a prayer with me. This was an experience from hell. My husband and I said a prayer and deep down I just knew something was not right.

FEMALE, AGE THIRTY-THREE, SEAT 12G

My subconscious told me, "Don't get on that plane."

The pilot began to taxi out, receiving his OK for the takeoff. The plane was sluggish on getting speed. When the plane got speed on that runway, it was bumpy, then I heard a pop and another pop, and then the pilot applied brakes. I got out of my seat and ran to the nearest exit.

CAPTAIN ONEAL L. SISSON

The takeoff roll was conducted within normal parameters prior to reaching our V1 speed of approximately 155 [knots]. Simultaneous with, or shortly after, the first officer's call of "V1," I observed a Door-open light. . . . I then announced to the crew that I was going to continue the takeoff by saying, "I'm going," or words very similar to that.

No more than a second or two passed before conditions changed significantly. I do not recall hearing the first officer make a "V1" call. I did not rotate the aircraft or otherwise pull back on the [control] yoke. Instead, I felt the aircraft vibrate and shudder violently and yaw to the left. I also perceived that the left side was settling in a lower position than previously held. I heard some popping sounds. I did not believe that it would be safe to attempt to lift the aircraft off the ground. Accordingly, I initiated an aborted takeoff.

I slowed the aircraft using full brakes and reverse thrust. The yaw was corrected by the use of brake and rudder inputs. First Officer Zeitz also applied brakes during approximately the last two thousand feet [of runway before stopping].

As we slowed, First Officer Zeitz reported our condition to the tower. Second Officer Cerulli mentioned that the brakes would likely be very hot and suggested that I consider an evacuation. When the aircraft came to a stop, I called for the evacuation checklist. I made this decision because I thought the potential for fire was significant. In addition, I heard a voice from the back of the aircraft say the word "fire."

MALE, AGE SIXTY-SIX, SEAT 3D

The plane . . . violently shook with an explosion-like noise at a ground speed of 155 to 160 m.p.h. And suddenly the brakes were applied and apparently the engines reversed and an explosion rocked the aircraft and it shuddered in an attempt to stop. I knew we would end up in the ocean. I held my wife's arm...and told her we'd be in the water. The flight attendant became hysterical and froze when the plane stopped and the fire was apparent. After she opened the door after a frantic order from [one of the passengers], she then panic-screamed, "Get out now!!" scaring the wits out of us. The flight attendant appeared petrified. She hesitated opening the door.

FEMALE, AGE THIRTY-SIX, SEAT 4F

I was sitting in first class, third row, and as I turned around to take pictures, I saw sparks coming from the engine. I turned to see it fully, and I saw flickers of fire. I screamed to others in first class that I think the engine is on fire. Someone sitting across the plane near the window asked, "What did she say?" At that time, as I got ready to repeat it, the plane was in the air at about two or three feet [and] came plowing down. I heard two loud explosion noises. Later, I was told that the noise was the wheels of the plane exploding because of the intensity that the plane had when it came down. I said, "We need to get off this plane," but it was still rolling. Once it stopped, the stewardess opened the doors and said, "Get out." She forgot to open the inflated slide. Someone said, "Open the slide." The man sitting next to me panicked. He screamed, "The plane is not going to stop. The plane is not going to stop." I told him it would and [to] calm down. The children in first class were crying and screaming. Once the plane stopped, people were trampling one another to exit the plane.

FEMALE, AGE THIRTY-FOUR, SEAT 16J

. . . I heard a loud BOOM! I look up and I see debris from the top of the plane falling. I then turn to the right of me and I see fire. I thought the plane was going to explode. I look behind me and I see the flight attendants and, to me, they all panic, screaming to run for your lives, save your life! Everyone then starts to scream and push. I finally get out. My husband let me off before him and I get to the chute and went down. I was in shock, and I just get to the bottom of the plane and take off running. I later stop and I see the fire trucks there at the plane. I reunited with my husband. I felt [that] only by the grace of God that [our] lives were spared.

MALE, AGE FORTY-SIX, SEAT 45F

I was shocked to see people jump out of the plane with so many personal belongings. My God! . . . People were jumping out holding cameras, overnight bags. I was shocked to see that. From my point of view, get the hell out of the plane and worry about your luggage later.

11. HONOLULU, HAWAII

February 24, 1989

United Airlines Flight 811

United Airlines flight 811, from Los Angeles to Sydney, Australia, experienced an explosive decompression as it was climbing between twenty-two thousand and twenty-three thousand feet after taking off from Honolulu, Hawaii, en route to Auckland and Sydney. The time was nearly two in the morning. The airplane made a successful emergency landing at Honolulu and the surviving crew and passengers evacuated the airplane. Examination of the airplane revealed that the forward lower-lobe cargo door had separated in flight and had caused extensive damage to the fuselage and cabin structure adjacent to the door. Nine of the 356 passengers were thrown through the fuselage opening and were lost at sea.

While climbing from twenty-two thousand to twenty-three thousand feet at an air speed of three hundred knots to the assigned flight level of twenty-eight thousand feet, the flight crew in the cockpit heard a "thump." An improperly latched forward cargo door had opened, creating an explosive decompression not unlike the explosion of a bomb. Indeed, the cockpit crew believed that a bomb had exploded. The airplane, a Boeing 747, rocked and yawed to the left. The cockpit filled with condensation. The cockpit crew put on their oxygen masks but no oxygen flowed through the system. They heard the "horn" activate and believed the passenger oxygen masks had deployed automatically. The captain throttled the airplane's engines to "idle" for an emergency descent and deployed the speed brakes. The landing gear remained up. He turned left to Honolulu. The first officer contacted Honolulu traffic control, declaring a Mayday. The captain shut down the number-three engine because of heavy vibration. The second officer began to dump fuel. The captain instructed him to check the passenger cabin. The second officer observed a large hole on the right side of the airplane, debris, and injured passengers who were putting on their life vests in preparation for a sea landing. The second officer also observed a fire in the number-four engine that he reported to the captain, who shut that engine down, and the fire appeared to extinguish itself.

At 2:19 A.M. the captain called Honolulu ATC. "We're missing a section of the right side of the airplane. Part of the fuselage is missing and we've lost number-three [and] we've got number-four shut down 'cause it appeared like we had a fire out there. We want all medical equipment we can get and all the equipment we can get standing by."

At sixteen thousand feet the captain slowed the airplane to 250 knots. He applied full power on engines number one and two and full left rudder trim. At four thousand feet he held altitude somewhat with a rate of descent of approximately fifty feet per minute.

At 2:30, the cockpit had the Honolulu airport in sight approximately ten miles distant. The cockpit crew extended flaps. Five seconds later, the Honolulu local controller cleared flight 811 to land on runway 8L.

The airplane touched down at a weight of 600,000 pounds and at a speed of 170 knots. The captain applied idle reverse on number-one and -two engines and gentle braking to stop the airplane on the runway. At 2:34, the cockpit advised the local controller that the passengers were being evacuated.

FEMALE, SEAT 3F

I recall being very aware of many details of the [preflight safety] video as never before, particularly how the stewardess put on her life vest while sitting down. I remember vividly imitating her moves as I put on my own life vest. I remember noticing, for the first time, how passengers had to jump to get onto the slide. I looked around for the nearest exits to my seat. As a frequent flyer I always look for the two exits closest to my seat. I also look at the size of the people I will be competing with to get out!

COCKPIT VOICE RECORDER

CAPT: What the [heck] was that?
FO: I don't know. [To Honolulu Control] Okay, it looks like we've lost number-three engine and we're descending rapidly, coming back.
HONOLULU: United eight eleven heavy, roger. Keep center advised.
CAPT: Call the aft flight attendant. Goin' down.
SO: We've lost number-three.

"B" ZONE FLIGHT ATTENDANT
[FROM THE FORWARD, OR B ZONE, OF THE AIRCRAFT]

We were setting up for snack/cocktail service. There was a loud pop and the cabin filled with fog. Debris seemed to fly everywhere. An elderly passenger was suddenly on the floor. . . . I looked around and did not see my oxygen masks hanging.

FEMALE, SEAT 3F

I had just released my seat belt in order to reach my purse under the seat ahead of me about five feet away. As I sat back down in my seat, the "bang" occurred. I fastened my seat belt once I realized I didn't have it on. At decompression, I turned quickly to look over my left shoulder to where the sound came from. I saw lots of pieces of "interior" flying around and lots of light yellow "dust" in the air. After I refastened my seat belt, I reached for my oxygen mask and put it on. There was no oxygen flowing, but I kept it on for a while to keep the "dust" out of my nose.

FLIGHT ATTENDANT NUMBER FOUR

Approximately [ten minutes after takeoff] I was about to walk from R2 jump seat area toward the forward right aisle [and] I was knocked off my feet. At the same time, I heard a very loud bang. Things [were] flying and I was grabbing for a solid object to hang on to. My right leg was caught under row 13GH, and after securing myself I looked up and saw a lot of structure damage. I felt that the opening was very close to where I was lying. So I hung on tight.

COCKPIT VOICE RECORDER

CAPT: Emergency descent.
FO [to Control]: United eight eleven heavy, we're doin' an emergency descent.
HONOLULU: United eight eleven heavy, roger.
FO: Put your mask on, Dave.
CAPT: Yeah. I can't get any oxygen.
SO: What do you want me to do now?
FO: You okay?
CAPT: Yeah.
FO: Are you gettin' oxygen? We're not getting any oxygen.
SO: No, I'm not getting oxygen either.
CAPT: I'm okay.
CONTROL: United eight eleven heavy, say your altitude now.
CAPT: Leaving fifteen [thousand feet].

Seconds go by.

FO: I think we blew a door, or something.
CAPT: Tell the flight attendants to get prepared for an evacuation. We don't have any fire indications.
SO: No, I don't have any [fire indications].
CAPT: Okay, we lost number three.

Seconds go by as the crew go through a shutdown checklist for the lost engines.

FO: Center, United eight eleven, you want to have the equipment standing by [and the] company [United Airlines] notified, please.
SO: Lots of fuel. Should we dump [fuel]? Want me to start dumping?
FO [to HONOLULU]: And Honolulu, United eight eleven heavy, we're gonna level at nine thousand here while we assess our problem. And we're coming back direct Honolulu.

HONOLULU: United eight eleven heavy, roger, keep the center advised.

FO: Okay.

SO: We got a hundred and eighty thousand pounds [of fuel].

CAPT: We got a control problem here.

FO: Do we?

FA: Everyone take your seats . . . take . . . everyone take your seats.

FO: Start dumping the fuel.

SO: I am dumping.

HONOLULU: United eight eleven heavy, when able forward the souls on board and fuel at landing.

FO: Okay, ah, stand by. We'll give it to you as quickly as possible.

HONOLULU: Roger.

CAPT: We've got a [heck] of [a] control problem here. I've got almost full rudder on this thing.

FO: You dumping as fast as you can?

SO: I'm dumping everything.

CAPT: We got a problem with number-four engine.

SO: Yeah, number four looks like it was out too.

FO: Well, we got number one.

SO: We're dumping five thousand pounds a minute.

FO: Ah, Center, United eight eleven heavy, ah, do you have a fix on us?

HONOLULU: Affirmative, sir. I have you on radar.

FO: Okay, it appears that we've lost number-three engine. Ah, we've lost . . . We're not getting full power out of number four. We're not able to hold an altitude right now. We're dumping fuel. So, ah, I think we're going to be able . . .

HONOLULU: United eight eleven heavy, roger. I show you six zero miles south of Honolulu at this time.

SO: I haven't talked to anybody yet [of the flight attendants]. I couldn't get to them. You want me to go downstairs and check?

CAPT: Yeah, let's see what's happening down there.

MALE, SEAT 15C

I was with my sales manager and friend and we were holding hands, saying our prayers, thinking we were going to die.

MALE, SEAT 27H

Most people were relatively calm. There was some yelling and some screaming but basically calm. There was one girl about five rows in front jumping around in the aisle, with blood on her. Quite hysterical.

"B" ZONE FLIGHT ATTENDANT

I began feeling light-headed and yelled [to another flight attendant] to get the two oxygen bottles out and operating. She appeared to be having a difficult time. . . . I was becoming quite light-headed. I ran to 5R but was so disoriented I could not remember where the [oxygen] bottle was for that [position]. I forced myself to remain calm and ran back to 4L. By then [another flight attendant] had two bottles going and shared her mask with me. Another flight attendant had the other bottle. The oxygen helped immediately. I crossed to 4R and told [another flight attendant] to get the other bottle going for me.

[One of the male flight attendants] had broken his arm. I got him ice. I put on the bottle and walked up into D Zone after checking to be sure [another male flight attendant] was OK. . . . He was. I got as far as C Zone and could see the damage forward was far worse. I walked back into D Zone and yelled, "Don your life vests and inflate one chamber." I got the red vest from 4R and yelled to the crew to put their vests on too. I continued into E Zone, shouting [to] the passengers to put on life vests. I got my blue and pink announcement books from my manual and opened to the checklist. I went to try the PA at 4R and it did not work. . . . [Then] I began to clear the galley floor for easier passage. . . . I instructed the passengers to review safety cards, locate nearest exits, and grab ankles. I removed heavy bags from overhead bins that were hanging open and told passengers to shove them under seats. I went back to E Zone with [a] megaphone and did the same thing. I had to help several passengers fasten the hooks on their life vests to the rings as they did not seem to understand how to put the vests on, or had inflated them before fastening the hooks. When it appeared that all passengers had vests on I told them to review safety cards and put heads down. I went to my jump seat and sat down.

FLIGHT ATTENDANT NUMBER FOUR

Seat 14H seat cushions were gone. No passenger [was] there. The passenger on 14G was right next to me, disoriented, and started to climb up her seat toward the back direction. I was hanging on to her, afraid that I might lose her. A male passenger came from the back and took her backward. I found the life vest under 14G, and I put it on with my left hand. My right hand was now locked around the armrest at seat 14G and H. My right hand was getting cold from the wind. I covered it with the blanket I was using for my legs. I could not see forward except to my left side and above me. Nothing much else to do but to wait for landing and to hang on. It seems [to take] forever to land. Passenger at 13D extended her arm to me and I touched her hand. Then she touched passenger 13G's hand. Passenger 14B had the oxygen mask on. The overhead ceiling in the business class was gone. The aircraft banked and I could see the lights. . . .

COCKPIT VOICE RECORDER

FO: What a hell of a thing to happen on your second-to-last month [before retirement].

CAPT: No shit.

FO: You got a fire out there.

CAPT: There's a fire out there?

FO: Yeah, it looks like it's engine number four.

CAPT: Which one?

FO: Looks like number four. Hold a second.

CAPT: Yeah, we got a fire in number four. Go through the procedure, shut down the engine.

Seconds pass.

CAPT [to Honolulu]: Ah, we're on two engines now.

SO [reporting back to the cockpit after an inspection of the cabin]: The whole right side. The right side is gone from about the 1R [seat] back. It's just open. You're just looking outside.

CAPT: Waddaya mean, pieces?

SO: Looks like a bomb. Yes, the fuselage, it's just open.

CAPT [to Honolulu]: Okay, it looks like we got a bomb that went off on the right side. The whole right side is gone.
FO: Anybody?
SO: Some people are probably gone. I don't know. . . .
CAPT: We got a real problem here.

Minutes pass.

HONOLULU: United eight eleven, roger, I have you on radar.
FO: We got forty-five miles to go.

"B" ZONE FLIGHT ATTENDANT

. . . The captain announced two minutes to landing. I ran to the back to get a pile of blankets for [the injured flight attendant]. He could not bend over. I sat down and belted in and began yelling for passengers to put their heads down, stay down. We continued yelling, "Stay down, heads down."

MALE, SEAT 27H

The pilot announced that upon stopping we would evacuate. On hearing this, our flight attendant gave myself and the guy in seat 27I instructions for evacuation, mainly waiting at the bottom of the chute to help others off.

The captain advised us that we would touch down in two minutes, and that we would evacuate. We could make out some instructions over the PA, but not very well. One was to put on our life jackets, but the noise made it hard to hear. My wife could not find her life jacket at first. It was being used by the cabin steward to show people how to put on their jackets. I noticed some oxygen masks did not drop. Ours did, but cabin pressure and oxygen-mask pressure felt no different.

COCKPIT VOICE RECORDER

FO [to Honolulu]: Okay, we have the airport [in sight] now, United eight eleven heavy.

HONOLULU: Eight eleven is cleared to land 8L, equipment standing by. Wind zero five zero, one two.
FO: Cleared to land 8 Left, United eight eleven heavy.
CAPT: Okay, well, let's try the [landing] gear.
FO: Okay, trying the gear.

Eight seconds go by.

FO: You're okay. You've got ten flaps. That should be one seventy but the inboards are up [and] the outboards are up, so, two hundred, one ninety's probably good speed.
SO: Two-engine approach.
FO: Two-engine approach.
SO: I hear people screaming back there.
FO: She's yelling for them to sit down.
CAPT: Give 'em [the] three-minute warning. Tell her on the PA.
FO: Are we depressurized?
SO: Yes. [To the cabin passengers and crew] Have about two minutes until we touch down. [To the captain] Want me to tell 'em to plan for evacuation?
CAPT: Yeah, when we evacuate.
SO [to cabin passengers and crew]: We will be evacuating. We will be evacuating upon touchdown and once we come to a stop.

Approximately one minute goes by.

FO: Lookin' good. One ninety-two. Thousand down. Dot and a half high. One ninety. One eighty-five. A little slow. A little bit slow, Dave, below what we want. One ninety. One nine one ninety-two. Comin' up to the glide slope. You got the gear down. We're cleared to land. Everything's done, as far as we know. Two hundred. One ninety-five. Half a dot high. Lookin' good. One ninety-two. One ninety-five.
CAPT: Comin' off the power.
FO: One hundred feet.
SO: Fifty feet.
CAPT: Center the trim, center the trim.
SO: Thirty. . . . Ten. . . .

Sound of touchdown.

SO: Zero.
FO: We're on.
SO: Gear's holding.
FO: No spoilers.
CAPT: I'm gonna go into reverse.
FO: On number two only, 'cause we got one seventy [knots]. One sixty. One thirty. One twenty.
SO: Looks good.
FO: One ten. Looks good so far. One hundred. Brake pressure's holding. . . .
CAPT: Let's go through the procedure.
FO: Brake pressure.

Eighteen seconds of rollout.

FO: Okay, we're stopping here, United eight eleven heavy. We're evacuating the airplane.
SO: I'm giving the evacuation signal.
HONOLULU: United eight eleven, roger. You got the airport.

"B" ZONE FLIGHT ATTENDANT

Because passengers kept looking up to see out the windows, we yelled throughout landing, rollout, and until the plane stopped.

When I knew we had stopped, I yelled, "Release your seat belts and get out." I turned on my Evacuation alarm. I did not see fire at my exit so I opened my door and yelled, "Stand back," as passengers were crowding the door area. . . . I pulled the manual handle [on the door] and said, "OK, you and you—to the bottom [of the slide] and stay there. Pull people off." I held my handle and yelled, "Come this way, form double lines." I could see [another flight attendant's] door was open and passengers were pouring out. The evacuation was very fast. I had to grab bags from several passengers, and went into D Zone to physically drag three passengers who were busy unloading the overhead bins. I had to pull the bags away from one man and shake him to get him to move. I saw a pilot in the cabin but there didn't appear to be any more passengers. I checked to be sure E Zone was empty and jumped into the door 4L slide.

MALE, SEAT 15C

[During the evacuation there was] panic. Some idiots were coming down the slides with duty-free goods and bags. Men were walking over women and children. I was disgusted. I find it very hard still to understand how people can walk over other people in times of an emergency.

MALE, SEAT 27H

. . . We were first off. I carried my wife clear first, and then went back. My wife was pregnant at the time. There wasn't really calm but not panic, either.

People just moved swiftly to get out.

12. DETROIT, MICHIGAN

December 3, 1990

Northwest Airlines Flight 299 and Northwest Airlines Flight 1482

Northwest Airlines flight 299, a Boeing 727, and Northwest Airlines flight 1482, a DC-9, collided near the intersection of runways 3-21 and 9-27 at Detroit's Metropolitan Wayne County Airport. The weather at the time of the mishap was one-quarter mile visibility, wind at eleven knots, temperature forty-one degrees Fahrenheit. There were no injuries aboard the Boeing 727, but on the DC-9, eight people died, including a flight attendant; twenty-one people, including the cockpit crew and the other flight attendant, were injured.

Flight 299 had pushed back from the gate at 1:31 P.M. The crew noted that the limited visibility was further deteriorating as they taxied to runway 3. As they taxied the crew observed a Northwest DC-9. "I lost sight of this aircraft as it taxied away from me. It appeared to be entering an area of lower visibility," said the captain of flight 299.

As flight 299 crossed Runway 9-27, the second officer commented that the weather was deteriorating significantly. The captain noted that he could see a distance of approximately eighteen hundred feet. Flight 299 stopped at the stop line on the runway and reported itself ready for takeoff.

Northwest flight 1482, meanwhile, pushed back from the gate and was cleared for Runway 3 Center. The captain reported that he had missed turning on taxiway Oscar 6, due to poor visibility. The ground control operators, notified of this missed turn, issued new directions and instructions to the cockpit crew. As the flight continued the taxi at a very slow speed, the first officer estimated that visibility had dropped to five hundred to six hundred feet. Ground control advised flight

1482 to report crossing runway 9-27. The captain stated later, "I stopped and could just see the beginning of a white line. Jim [the first officer] was talking to ground control, and I saw off to my left side what looked like a flashlight or a small diamond. I realized it was a white light, which told me I could be on an active runway. I taxied the airplane to the left of the runway edge and stopped. I picked up the mike and told ground control we do not know where we are, or we are lost, or something like that. I then looked up and saw the Boeing 727 coming right at us."

After receiving takeoff clearance, flight 299 rolled down runway 9-27, and the first officer called "Eighty knots." At approximately one hundred knots, "we entered an area of limited visibility [and] suddenly an aircraft appeared on the right side of the runway and facing us," said the first officer. "I shouted and flinched to the left and moved the yoke to the left and slightly aft, all simultaneously. I commenced abort procedures, max braking, thrust idle, speed brakes deployed, and reverse thrust."

The first officer of flight 1482 said, "I instinctively ducked over the left. The B-727 wingtip impacted the cockpit to my right. Although severely dazed I distinctly remember hearing [the captain] say, 'Evacuate the aircraft' three times over the PA. I recall thinking, 'Do the evacuation checklist,' but I can only actually remember pulling the two fire-shutoff [handles]. My seat was jammed and my right leg was caught between the control column and the instrument panel. When I finally reached the cockpit door there were about five or six people at the exit and a flight attendant on the ground screaming, 'Jump, jump, jump.' I recall pushing one reluctant passenger overboard and the others quickly followed. I then went back as far as the first class bulkhead, and at that point several rows in front of me and to my left I saw fire and some people in the aisle. I was almost immediately engulfed by a billowing black cloud of smoke." His eyes were burning and he could not breathe, so he returned to the main cabin door and exited the airplane. He and several passengers assisted an elderly lady off the wing. He stated that they were the last two people off the aircraft. He located [the lead flight attendant], who told him that [the other flight attendant] was still in the aircraft. [The first officer] attempted to climb the slide to reenter the aircraft but fell down and was subsequently restrained by a fireman.

MALE, FLIGHT 1482

... I only had probably no more than ten minutes to get from my plane arriving to my departing plane. I got there in Terminal D, I think it was, and I had to go all the way over to Terminal C. I was trying to move pretty fast and ran to try to make my connection, and I caught a cart, a motorized cart, going down the aisle and asked if I could ride. He said, "Yes." So I got over there just in time to catch the plane before they were finished loading. I was probably one of the last ones on. We got on the plane, and after it began to taxi out of the parking spot there, it just taxied on down the runway somewhere and then it came to a runway that I'm sure planes were using, and it stopped. There was a plane in front of us that appeared to turn left. I thought it turned left and went up that runway. We stayed there and in just a short period of time, I saw a little small jet. It looked like a private jet, come by, down that runway, and take off, then I saw a plane that was probably close to a DC-9—I'm not real good at identifying those—but it was about a DC-9 come down there and take off.

Shortly after, we turned left onto that runway, and it was pretty foggy out. I could hardly see the signs, the little small numbered signs on the right side of the runway. Apparently, that identifies what runway you're on or something. Anyway, we turned left on that and we were probably thirty or forty-five seconds up that, we turned around, and I assumed we were turning around to get our attitude for takeoff, so we'd be in the same direction as the other planes were.

... Anyway, I heard the engines surge a little bit, and I thought we were ready to take off.

ELDERLY FEMALE, SEAT UNKNOWN, FLIGHT 1482

I checked in, I mean, I had a seat assignment from the travel agency ... so I went when they called [for boarding of the flight], and I didn't realize that I would be in the very last place in the plane, and I thought to myself, "I'll have to tell my travel agent to get me up in the middle [of the cabin on a future flight]."

Anyhow, we were fiddling around there, fastening seat belts, looking at the magazine, and there was a young man across the aisle from me in the aisle seat, and he kept coughing, and so I gave him a cough drop from my purse. We didn't talk, and he just said, "Thank you," and popped it in his mouth, and I arranged my package that I had with me under the seat, and I was thinking of putting my purse under, too. Then I wondered if I should go into the aisle seat. I was in a row with two seats in the aisle, and I was in the window seat, so I thought maybe that this might be better to be out there, but I didn't move because the stewardess began making announcements. I really could not understand a word she said. [The volume on the PA system] was very low, and it was not clear. It came out kind of, well . . . somebody might say it was my hearing, but I've been on planes before and I knew what they told anyhow. So I looked around to see where the exits were, and I could see [them], and I noticed that back there where the toilets were, there was another Exit sign, so I thought when [the flight attendant] comes around to check the seat belts I'll ask her if it would be all right if I changed [seats].

. . . So, I looked out the window. I was still on that window side, and I thought to myself, "This is incredibly, not hazy or foggy, although foggy is part of it, with pale yellow fog." I thought, "Not gray fog." And I thought to myself lots of times when I would be listening to the radio or TV at home, they would say . . . "No flying out of Pitt now for four hours at least because visibility is so poor." And I thought, "How on earth can they get up beyond this [fog]?" I've never been in a plane that I could not see out of the window, [not] see anything. And I was thinking to myself, "Well, maybe they have equipment that helps the pilot get over this, above it, and then we'll be on our way."

MALE, SEAT 23D, FLIGHT 299

Our plane taxied immediately out to the runway. We looked out. We have about three-eighths-mile visibility by my reckoning. By the time we turned onto the runway I could see the white lights. I said to my seatmate [Seat 23G], "I doubt if we have five hundred feet of runway visual range on this runway." The pilot brought his thrusters right up to takeoff power and we went down the runway. . . .

MALE, SEAT 1B, FLIGHT 1482

During the taxi, the aircraft was passing through small patches of fog, stopping and continuing several times. At the point before we made the final turn, the fog was extremely intense. I could see the wingtip and an unlit marker light on the left side of the aircraft. Looking down I saw the white bar lines of the runway directly under my window as we crossed down. I found this to be unusual because the cockpit door was still open and the flight attendant was still in the cockpit. The aircraft again was stopped at this time. I could see both the flight attendant and the first officer looking around out all the windows very alertly as if they were looking for other traffic. I could hear the radio traffic but was unable to decipher the words.

FEMALE, SEAT 14D, FLIGHT 1482

I looked out and I was a little apprehensive, I have to say, because it was *very* foggy. I couldn't see a thing. We started taxiing along the runway pretty slow for a little while, a few minutes, I guess, and the guy in front of me, he said, "I'm glad he's driving."

MALE, SEAT 9G, FLIGHT 1482

I remember looking outside. The fog was extremely [thick], and I kept telling myself that there was no sense in being concerned. They're experts. This is what they do. I guess if there is anything I was concerned about it was the fog and the visibility, but . . . you fly through clouds, you fly through things, you know? They fly by radar or whatever it is. That was the only concern. . . . It was a constant rush all day. I was kind of relieved when I sat down and we taxied on time. It was only a minute or two late that we left the gate. We taxied for a good period of time, and I wasn't really paying a whole lot of attention. The only thing that would get my attention was the fog from time to time.

COCKPIT VOICE RECORDER [FLIGHT 299]

The cabin crew members talk about their seniority status, their careers, and possible career moves, then push back from the gate and start to taxi.

CAPT: What was the last visibility they gave . . . ?
SO: Three-quarters of a mile, he gave.

Three minutes go by, and the crew receives weather briefing as they taxi.

FO: It's going dog shit in a hurry, isn't it?
CAPT: Yeah, boy. I don't see how it's even a quarter of a mile [of visibility].
FO: I don't either.
SO: Not in here it isn't.

Another three minutes go by as the crew starts the number-three engine, then goes through its takeoff checklist.

CAPT: Are they ready in the back there?
SO: Okay. Panel items complete. Allowable takeoff weight checked. Probe heat?
FO: On. Lights out. [Seconds later] Takeoff numbers checked, set.
SO: Flaps?
SO: Fifteen, fifteen blue.
SO: Trim?
FO: Zero, zero, five point nine.
SO: Controls?
FO: Controls?
CAPT: Free and normal. [Two or three seconds go by.] Tell him we're ready to go.
FO: Okay.
SO [on PA]: Good afternoon, ladies and gentlemen, from the front cockpit, welcome aboard flight two ninety-nine to Memphis. We're currently number one for departure. We should be airborne fairly shortly. Flight attendants, please be seated.

COCKPIT VOICE RECORDER
[FLIGHT 1482, WAITING FOR PUSH BACK FROM GATE]

FO [to CAPT]: I got shot down once over in Southeast Asia and ah . . .
CAPT: Oh, is that right?
FO: I didn't have time to get scared.
CAPT: Yeah.

FO: And then, ah, when I was flylin' T-38s one time, I had a fire, an engine fire. That was a simple procedure in that airplane because, ah, if they, if the fire was confirmed, bold face was: throttles closed, engine-fire-shutoff switch pull, if fire is confirmed, eject. And you could confirm it, you know, with rough EGT high, or high EGT, or fire lights, and in my case, the tower controller said my call sign that day was DAY-21—"DAY-21, you are on fire. Eject!" So my decision was made. Bam! I . . . .

CAPT: Was this right after takeoff, or something?

FO: Right after takeoff, yeah.

CAPT: Wow.

FO: After a touch-and-go.

CAPT: Wow.

FO: It turned out what had happened was that we sucked a bird up in there and that blew the engine up and then, ah, somehow a fuel line got cut.

The captain and first officer discuss the details of FAA drug tests, their choices of airlines to work for, and pay, and other nonpertinent conversation.

CAPT: That fog is pretty bad here.

FO: I'll be brokenhearted if we don't get back. [Several seconds go by.] Guess we turn left here.

CAPT: Left turn or right turn?

FO: Yeah, well, this is the inner here. We're still goin' for Oscar.

CAPT: So a left turn.

FO: Near as I can tell. Man, I can't see shit out here.

CAPT: Yeah.

FO: Man, this is . . .

Several seconds go by, as flight 1482 tries to find out where exactly on the taxiway it is located.

CAPT: So what's he want us to do here?

FO: You can make the right turn, he said, and report crossing [runway] 27 and then I'll ask him.

CAPT: This . . . this is a right turn here, Jim?

FO: That's the runway.

CAPT: Okay, we're going right over here then?
FO: Yeah, that way. [A half minute goes by.] Well, wait a minute. Oh, shit, this is, uh, uh . . .

Another thirty seconds go by.

CAPT: When I cross this, which way do I go? Right?
FO: Yeah.
CAPT: This is the active runway here, isn't it?
FO: This is, should be, nine and two seven. It is . . .

Several more seconds go by.

CAPT: Give him a call and tell him that we can't see nothin' out here. [Several more seconds go by.] Tell him we're out here. We're stuck.

End of recording.

COCKPIT VOICE RECORDER [FLIGHT 299]

Crew goes through final takeoff checklist items.

FO: Boy, this is dog shit now.
CAPT: Yup.

Sounds of increasing engine noise as takeoff roll begins.

FO: Eighty knots.
COCKPIT: Oh. Oh.

Sounds of crash.

CAPT: Abort.

FEMALE, AN OFF-DUTY FLIGHT ATTENDANT, SEAT 1D, FLIGHT 1482

I looked out and I said, "You know, it's really weird, but I think we're sitting on the runway. Why are we sitting on the runway and not going anywhere?" Usually before the pilots take off or even before you get on the runway, they tell [flight attendants] to sit down, and I thought, "Well, he never told anybody to sit down," and then I thought, "Well, maybe we're not on a runway but it looked like the runway to me." I don't even know why I thought that. And then the guy said, "Well, maybe he's trying to decide whether to take off or not with all the fog." And he said, "Maybe he can't see." And I just figured that's what was happening when we were waiting to see if we should take off or not, and then all of a sudden, boom!

FIRST OFFICER, FLIGHT 1482

When I saw the other aircraft coming at us, to my way of thinking time slowed down into, I don't know, it seemed like it took him forty-five seconds to travel about six hundred or four hundred feet or however far away he was. And I remember I said, "Duck!" and leaned over the center console. I don't know why, just like trying to dodge a bullet, I guess.

MALE, SEAT 23D, FLIGHT 299

I knew we were heavy. We rolled for a long ways. And just before the nose was supposed to go up, I heard a big bang right underneath. We were over the wing. The pilot immediately went on the reverse thrusters and brakes and I looked out on the right side and saw we had lost the right outboard wing section.

MALE, FLIGHT 1482

The plane began to move and about that time I heard this loud noise and explosion and concussion go by my head. I got hit in the side of the head with something. I don't know if it was glass or debris or from the walls or what, but the explosion and bright light seemed to come from my right rear quarter. Somebody up front yelled that an engine exploded, so I relaxed for a second, and I thought, "Well, what hit me in the head?" So I started feeling my head and my ears and my hair to make sure everything was intact. Then I realized that there was smoke. So I unbuckled my seat belt.

COCKPIT VOICE RECORDER [FLIGHT 299]

FA: Ladies and gentlemen, please remain seated. Ladies and gentlemen, please remain seated.
SO: Evacuate?
FO: Ah, do you want to . . . ?
CAPT: I don't think we'll need to, if there's no fire.
SO: There's no fire or anything, right?
COCKPIT: No, no.
SO: Okay, everybody stay seated back there, for now.
FO: Ah, great.
PA: Ladies and gentlemen, please stay seated.
FO: Shall we, ah . . .
SO: We were cleared for takeoff, weren't we?
CAPT: Yeah, and they even cleared the guy behind us into position and hold.
SO: We're okay. How're they?
CAPT: Is, is, is there, ah, is there, ah, anything going on? Stick your head out that window and look at that wing.
FO: I can't see any . . .
SO: Want to get the people off the airplane?
CAPT: Check the wing. I think it's mi—

End of recording.

MALE, SEAT 9G, FLIGHT 1482

And then all of a sudden there was a tremendous crash on the right-hand side of the plane. I could feel it. . . . Then I remember turning my head and ducking out of the way, and as I turned back to look back, I could see a ball of flame. I could see just orange balls on the right side of the plane. I can't remember if it took up the whole side but I could see a lot of [flame].

FEMALE, AN OFF-DUTY FLIGHT ATTENDANT, SEAT 1D, FLIGHT 1482

I don't know if I saw [flight 299] because we were looking out the window and as soon as it hit, I thought, "Oh, my gosh, that plane hit us," so I don't know. Maybe I did see it. And now I don't remember seeing it, but in my mind I saw it, because why else would I have known what . . . happened?

Then you just saw everybody screaming, and the guy that was sitting right beside me kind of grabbed me, and then I think he knew that I was all right because I was moving and the first thing I thought was, "This is it. It's gonna blow. We're done." And I thought, "Oh, my gosh."

ELDERLY FEMALE, SEAT UNKNOWN, FLIGHT 1482

And just about then, this terrible crash came. It's hard to know what was happening, except I think it was a big piece of plastic, maybe two to three feet wide and about three or four feet long at least, cracked probably off the roof of the plane, and glass began to shake, and I looked across at this young man to whom I had given the cough drop, and I could see flames a foot and a half at least high along the edge of the lower plane section, below the windows. I moved so fast I didn't have time to see if there were any [flames] on my side of the plane. But I thought, "There's an exit right behind me and I must get to it."

FIRST OFFICER, FLIGHT 1482

Anyway, I leaned over the center console, heard the impact, heard and felt the impact, and then I distinctly remember I heard the captain say, "Evacuate the aircraft." . . . I found out later, like two days later, that he had gone out the window, and I was not aware of that fact, because of the dazed state that I was in.

FEMALE, SEAT 14D, FLIGHT 1482

. . . Something hit me right here and someone pulled me out of my seat. . . . I don't remember taking my seat belt off and [a girl passenger] started pulling me. She said, "You gotta get off. You have to get off."

MALE, SEAT 23D, FLIGHT 299

The captain, when he came on the speaker, said, "We were cleared for takeoff, but we hit another airplane." And that was after we've stopped. He sounded very calm. He said, "Everyone stay in your seats." I could see the fuel pouring out of the right wing [or] what was left of it. It was going down the runway. [The pilot] kept the plane right in the center. The fuel ran down the right side of the runway.

FEMALE, AN OFF-DUTY FLIGHT ATTENDANT, SEAT 1D, FLIGHT 1482

Then I saw everybody just start running up front, and then I turned around and looked in the back and all of a sudden you could see fire started already, and the black smoke started coming, and that [was] when I knew, it wasn't just that we were hit and you could slowly get out. We were on fire and we *had* to get out. And I almost went to grab my bag because my life is in my bag, and I carry everything in my Northwest bag, and I thought, "I need this," and I turned around and looked at the lady in 2D, and she wasn't moving. She was just sitting there holding her head. And the first thing that went through my mind was, "That lady's not going to get out. We're on fire. I gotta go get her." I don't ever remember getting up. I don't know how I got back to 2D, because everybody was running forward, but I remember going back to her saying, "I'm another flight attendant. I can help you."

She just said, "Please don't leave me. I'm scared. Help me." And I said, "I will."

I got her up, and I think she was cut in the head and something had hit her but I don't remember, and I ended up getting up. I don't ever remember walking up to the door. I don't ever remember looking in the galley. I can't remember that side at all, and we went to go up there, and the passenger door wouldn't open. They couldn't get it open. It was only open about . . . not even a foot wide. And I thought, "Why isn't this door opening?" And I'm pretty sure the first officer was standing there, and he's the one [who] pushed the door open, and then all of a sudden everybody just started jumping out. They were like going crazy. And everybody just started jumping, and . . . I saw the first one fall down, and she like almost splatted, and I thought, "We can't have people jumping out of here like crazy," and I saw the flight attendant, the lead flight attendant, was already out. She was down, and she was at the bottom screaming, "Jump, jump," and I thought, "These people were going to get hurt jumping." I remember looking down and looking at the slide pack thinking . . . "Why isn't the slide working, why didn't it work?" And it never went through my head that . . . you have to pull it and that . . . it's broke. I thought, "Well, it doesn't matter. It's not that high. I would rather have some broken bones jumping than somebody burned and kept inside."

FIRST OFFICER, FLIGHT 1482

When I sat back up, the thing that came to mind was, I have to do the evacuation checklist in the absence of the captain. To the best of my ability, I did that. However, the only thing that I could swear to, actually . . . say that I did was pull the two fire-shutoff handles. I may have done other things, but I cannot say that I did those. Then I got up and went out the cockpit entrance door. There were . . . five or six people still waiting to evacuate the aircraft out of the left forward entrance door. One lady in particular was reluctant to leave the airplane. We helped her out and the other four or five people left in quick succession.

MALE, SEAT 1B, FLIGHT 1482

I do not remember unfastening my seat belt. By the time I stood up, there were people in the aisle. I attempted to go to the rear of the aircraft. Then I saw a person lying on the floor in front of the entrance door [who had] apparently tripped or [been] pushed. I yelled out something like, "Don't push, move forward." At this time, I could see the aisle full of people standing. The line was not moving, and I could see a wall of smoke building up at the rear of the aircraft. There was no one in the galley and I exited through the galley door. I do not recall opening the galley door, nor do I remember anyone else going through before me. In the second or so it took me to reach the galley door, the smoke had become so intensely thick [that] I could not see the rear of the aircraft any longer. As I exited the galley service door, the chute was not deployed and I jumped to the ground.

FEMALE, SEAT 14D, FLIGHT 1482

I think I'm in shock from the pain because I think I was hit with something flying through the air, and [an unidentified woman] steered me through here somewhere, and she said, "You're gonna have to get off. Jump down." I'm looking. I said, "There's no slide." She said, "You're gonna have to jump." I'm saying to myself, "God, that's real high." Now, this is all within like two seconds that all this went through my mind, so I did jump.

FEMALE, AN OFF-DUTY FLIGHT ATTENDANT, SEAT 1D, FLIGHT 1482

. . . I looked back [the length of the interior of the cabin from the door where she was standing] and I saw the black smoke and I thought, "There's people back there." I thought, "If I go back there I'm not coming back, but I don't even know where they are because I got on [last, and] I didn't know where people were even sitting in the back cabin." I thought, "If I walked back there I don't know if . . . there is [anything] I can do." I looked [down] at the ground thinking, "I don't want to jump." It was high, and I could see why people were scared to jump, and I thought, "Well, just get out," and I remember standing there trying to get the people out, thinking, "Am I going to have time to jump or am I just going to be blown out?" I didn't know how long we have, if it would blow up, and I thought, "I'm supposed to know what I'm doing. I gotta stay there and make sure the people were off, because I knew the lead [flight attendant] was already on the ground. They were gone. Everybody else was gone. And then I jumped off, and we got people out, and we tried to drag them away from the fire. . . .

FIRST OFFICER, FLIGHT 1482

. . . At that point I went back toward the first-class bulkhead. To the best of my recollection, there was not . . . there was fire and there was smoke, but there was not the black encompassing smoke that they encountered shortly thereafter. That came from the rear of the airplane and I don't know what caused [it] to happen. But, it was like a billowing black cloud of smoke. And I proceeded at that time back to the left forward entrance door and jumped out.

. . . Then I talked to [Flight Attendant] Anne [Hanson], and she said, "[Flight Attendant] Heidi [Joost] is still on the airplane." And I tried to go back up into the airplane and see if I could get her and I could not do that. And so then I assisted people to the best of my ability until I got too cold to do anything.

MALE, FLIGHT 1482

I looked at the guy next to me. He appeared to be unbuckling his seat belt, so I figured he was all right and ready to go. So I got up and got into the aisle and the crowd was surging forward. I began to think that there ought to be a door back here someplace. I remember seeing one, and so I looked on the right side. I didn't see it. I looked on the left side, and I saw about four seats back, there was a door and it wasn't open. So I yelled for somebody, "Hey, there's a door. There's an exit door here." And somebody opened it and people started going out that door.

About that time, I began to get blind in my right eye, and my left eye didn't have a lot of vision to it, but I could see the door, and I was about four seats away and I just held on to the seat and stayed with the people in front of me. I stayed against them by touching them and until I got to the door. By that time, the fire was coming up pretty close to where I was. I could really feel the heat and flames. . . . I put my hands on the door frame because I couldn't see that well, and I began to lift my leg up over the frame, the bottom frame of the exit. . . . And I got my knee and the surge of the crowd behind me pushed me completely out the door, so I went flying out the door.

My knee hit against the framework on the bottom and then when I got outside I landed on the airplane wing. I think I landed on my knee and hand. There was somebody down on the ground saying, "Jump, jump." So I jumped off the wing of the plane onto the runway, which was concrete.

MALE, SEAT 9G, FLIGHT 1482

I got out of my seat belt, followed a good many [other passengers] up front. There was a lot of chaos and a lot of commotion and people screaming and I guess I was about ten deep in line. There were several people in front of me. . . . When I walked from my seat, the ninth row from the front . . . the visibility within the plane was very good. I could see everybody in front of me when I got to the plane's [door]. Within what seemed to be seconds, you could smell the smoke, and the smoke started thickening in the plane. I took about three or four breaths of smoke, and I couldn't see any light. Everybody was pretty much at a standstill, and then screaming. I thought at that time, something went through my mind, I thought because of the smoke and having breathed it, I had taken . . . I don't know why, for some reason, the third breath of smoke, I could actually feel it, and it started to thicken inside the plane, and I don't know what happened at that time. People scrambled. Everybody moved around, and I ended up in a position where I could see the door by the front of the plane open, about a crack. I went up to the door, and I saw daylight and I pulled the door back and I jumped.

MALE, FLIGHT 1482

. . . So, as I'm coming back to the exit door there's other people still coming from the tail, and we were kind of coming in the same spot together, and I started yelling, "Let's take turns here. There's plenty of time, really, to get out, so let's take turns. Don't knock one another over." And amazingly, everybody—they might have been yelling, "Let's get out," and this and that—but nobody was . . . going crazy, and people did pretty much . . . take turns. You know, nobody ended up fighting at the door or something like that. Nobody really held [the line] up. Everybody went out—boom, boom, boom. You know, one-at-a-time type [of] thing? . . . I thought, "I may not get my turn," because this ball of flame I'm looking at pretty much is just coming on up [toward me]. You can see the seats kind of disappearing as it's coming forward.

And so when I was able to exit out the door, the first thing I saw is one fellow crawling away from the plane. . . . And the fellow ahead of me went off the wing and he just ran right off the wing like he was running and jumping into a swimming pool. I thought, "Oh, my God." And he hurt his leg when he hit, and he was hobbling.

So, anyway, I decided [I'd run] to the edge of the front end of the wing . . . and sat down and slid off. I've had two back surgeries . . . so I'm probably more conscientious of jumping like that. But anyway, [I] slid off and dropped to the ground, landed on my feet, and everything felt fine at that point. . . .

ELDERLY FEMALE, SEAT UNKNOWN, FLIGHT 1482

I couldn't see when I stood. I unfastened the seat belt, and I looked, but I could not see [the exit], and I thought, "I can't spend time hunting it. I must get to the other [exit]." Oh, and when the crash came, the young man jumped out of his seat as if he were catapulted, and he said something which I'm not clear about, but it was something to the effect, "Oh, my God," something like that. And I hardly had time to think. I didn't think [about] what had happened, but I just knew something terrible had happened, and you should get away from the fire. So I began to plow my way back up that aisle which was filled with debris, and I don't know when I went so fast. I thought, "I'm going to be all cut up with the tramping." Along the right side of the plane was a man seated in the aisle seat, and he still had his seat belt on, and he had his jacket on, and he had slumped over into the aisle with the top part of his body, and I knew he should get out, but I knew I couldn't do anything about it. He looked too heavy for me to do anything about it. But it went through my mind that he should be alerted. I don't know what had happened to him, but I . . . pushed his shoulder toward the window so I could get by, and I felt bad about that, that I couldn't help him.

Then, by that time, I was up in the area where I could look out to the left and see the open window and knew that's what I was headed for, but between me and there, there were a lot of flames and smoke, and I thought, "We mustn't go into that."

And two men I could see out beyond the open window standing on the ground by the wing . . . kept calling to me. I can't remember the exact words of encouragement [that the men used]. "Come on, don't wait, get out, hurry." You know, things [like that]. So all this time

I had my purse in my hand. [I] didn't realize I did. But I knew I couldn't get out that window without using my hands, because it was quite a step up, so I threw the purse out on the ground and . . . and then they kept calling to me to come, and the trouble was, it wasn't a tight skirt but it was mid-calf and it didn't have any pleats, so it didn't give as much as it might have if I had worn something else.

But I got both hands around the edges of that window and got out on the wing, and I remember thinking [that] when I got to the window there [should have been] a chute, but there wasn't anything. It was just the wing. By lifting my skirt up high I could move my legs enough to get out over the bottom part of the window, and I just stood there, and they said, "Come slowly, keep coming, we'll have you," and they reached up and each one helped me from one side, and I think both [men] took my arms, and the next thing I knew I was standing on the ground between them.

MALE, FLIGHT 1482

. . . I started to take off to run, and I looked back, and I saw a pregnant woman coming out of this [exit] doorway up there [in the airplane]. I thought, "Oh, my God." You know? So I started running on back, and I looked over to the side and I saw one other fellow [who] was kind of between me and up toward the front hatch, where I saw some people jumping, and I gave them a call to come on down and give me a hand, so he did and he got down there just in time, when this pregnant woman got to the edge of the [wing]. We told her to sit down and slide off, and she did, and we caught her.

I think we caught about two more people, maybe a woman and a man, and then this fellow comes out of the [exit] door and he's all fire. The very front of him isn't on fire, but his whole back and his arms, legs, [and] everything is burning. So, he came running out and sat down and slid right off, and we caught him too and kind of wrestled him down to the ground. We were yelling, "Get to the ground." And he pretty much did roll to the ground, and we got him down, and the pavement's wet there. We were on the cement, not on the grass at all. [We] got [the fire] out. [It] looked like he was cut, so the other guy and I kind of let loose of him a little bit and boy, he jumped up to get going, and his clothes were ignited . . . started to burn again.

So we got him back down on the ground and held him there, but we couldn't get [to] the fire around his neck. . . . The flames were coming up on both sides of his neck around his ears and that kind of stuff, so I reached with both my hands and put it out with my hands, and that's how I burned my hands, really. So . . . he got up and got going and in the meantime I'm looking down the other way and I saw people . . . jumping out of there yet. Somewhere right about that time, I looked back and I didn't think there was anybody else coming, but there was a woman standing in this [exit] doorway. . . . And she's just standing there. So, I go over to the wing, and I'm yelling at her to come on, and she's pretty well all sooted up black, and her face is all red from the heat, and I can't see her back yet, but when she finally did come out her whole back had been burning but the fire seemed kind of out. But, anyway, we had to coax her and call her and she finally stepped out of that thing with one foot and then finally came out with the other [foot] and came on the wing and just stood there. And we just couldn't get her to jump. So I ended up, I said, "Sit down and just slide off," and she wouldn't do nothing. I says, "Please, come on," and I'm pleading with her. It's like trying to get a kid to jump into a swimming pool for the first time and [I was] saying things like, "Trust me. I promise I'll catch you."

So, finally she sat down and slid off, and this other guy and I caught her. So, anyway, we were walking her out into the field there, and I was talking with her, and she seemed fairly lucid. She was talking to me and saying things like, "Boy, this is too much for an eighty-two-year-old woman." I says, "You better believe it. It's too much for a twenty-two-year-old." So we chatted a bit, kind of walking out, small talk, and I says, "I can't believe you got to the doorway because the guy just ahead of you was all burning, and I can't believe you made it, because it was a little later [that you exited]." She said, "Well, I was stepping over people coming out to get to the door."

MALE, SEAT 1B, FLIGHT 1482

I then observed the right engine on the ground enflamed. The right pylon area of the aircraft was also enflamed. The right wing was not burning. I ran around to the front left side of the aircraft. I believe at this point I saw the captain exiting through the sliding [cockpit] window using the emergency rope. The first officer was standing at the entrance door attempting to discharge the emergency chute. I believe the chute was in his hands. There was another person standing under the door beside me. I recall shouting, "Jump, just jump." . . . People seemed to exit very orderly through the door. We assisted people to the grassy area. The people . . . exited in a very short period of time. When no one else was exiting the aircraft, I then believed that everyone was out. I went over to the crew, who decided at this time we should move the people further back, and we did. I went to the flight attendant who had just moved the last person back and said, "It's OK. I think everybody is out." She replied, "The other flight attendant is not out here."

The aircraft passenger compartment was now fully engulfed in flames. I took my sweater off and made [the flight attendant] put it on.

FEMALE, SEAT 14D, FLIGHT 1482

I was stunned because it was cement [that I landed on after jumping off the wing], and I think that's when I might have fractured my rib, so I couldn't jump up right away, of course. I laid there for a while, and some guy came up to me and started pulling me away from the plane. He said, "It might blow up, honey. You're gonna have to get out of here." He grabbed me and started [to] drag [me]. I said, "I can't walk, I can't walk." He lifted me and carried me, I don't know which way because we got lost. He carried me across the field. He says, "I don't know where I'm going. I can't see a thing." He said, "I think we're going the wrong way," and we turned around and started going another way.

FEMALE, AN OFF-DUTY FLIGHT ATTENDANT, SEAT 1D, FLIGHT 1482

. . . I remember the lady . . . I got from 2D. She was really scared. She was just saying, "Please don't leave me." I knew she was hurt, but she would be all right, and they kind of start dragging her off of the cement to get her away from [the aircraft], and then I went over trying to see how many people were out, and I ran to the lead [flight attendant]. I said, "Where is the other flight attendant?" She said, "She's not here. She didn't get off. She's still in there." But there was nothing you could do. There was no way we could go back in, and I knew if we went back in [we wouldn't] come back out, and I knew everybody wasn't off the plane because there weren't enough people out there, so we're trying to walk around to see who was [there] and to get people further away, and I remember just seeing some man lying flat on his back . . . in the snow, and . . . and his face was totally black, and I went over to him, but you know, this guy needs help, but the lady [who] didn't want me to leave her, she's all right, and I can't remember if I said, "Somebody, would you just stay with her?" But I remember leaving and going to this guy, who said, "Would you please just help me up, get me out?" And I thought, "I shouldn't move him because he could have hurt his back," but he really wanted [to get] up, and I said, "If you start to hurt at all, go ahead and just tell me and we'll kind of drag you." But he was still too close to the plane. If it blew, I knew we were going to be in trouble, so I got him to stand up, and the first thing I said was to the other guy, "I don't want to touch him." He was totally burnt. I've never seen a white man's face totally black, and I still remember his eyes. They were like totally white, you know. You could see there was just like kind of glass eyes, and I grabbed his hands and it's like hunks of skin were already off, and then I went to look at his back. . . . There was nothing on his back. His shirt, everything, was just totally burned, but he could walk. His legs were fine. And I said, "Are you sure you can walk?" And he's like, "Yeah, I can walk." So he got up, and . . . we walked him a ways, and I just kept standing there talking to him, and I . . . asked him his name. And I said, "Gary, it might hurt to talk but just keep talking to me and you won't realize what happened to you." . . . I just figured he was going to go into shock, because I know if that [had] been me and I was burnt and I . . . saw what happened, I would probably start going bananas, but I knew I was fine. I wasn't cut or anything, so I just kept talking to him, and I asked him where he was from and he was from Morgantown, and he was in between jobs, and I said, "Does it hurt to talk?" And he said, "Yes." And I said, "Well, at least if it's hurting to talk, you're thinking about how it hurts to talk and nothing else," I said, "so just keep talking to me. I don't know what you can talk to me about, but just keep talking," I said, "that way you won't realize what's happening to us." . . . So I sat there and talked to him, and then another guy was just standing there, and I tried to say, "Why don't you just come over here and talk with us?" 'Cause he had a great big gash on his head, and he said, "I don't know what's wrong. I just see blood running down my face," and I said, "Well, let me see it." And it was a pretty big gash, and I think he felt a little bit better knowing what was wrong with him [rather than] just walking around knowing he was cut and bleeding.

MALE, FLIGHT 1482

The best I remember, I walked probably fifty feet or more to get to the end of the runway and then into the slushy snow of the grass that was to the other side of the runway. . . . I walked in a daze for a little bit, and then I decided that I better put something on my eye, so I pulled my handkerchief out and took my glasses off and put my handkerchief over my eye.

The only thing I heard, I thought it was the captain [who was already out of the airplane], and he was saying, "Get out of the way. Move out, because the plane may blow." There was another guy down on the ground by the wing, and he was saying, "Jump, jump, I'll help catch you." And he was a volunteer. He was a passenger.

. . . We must have all stood around there. It seemed like a long time before anybody ever came there.

MALE, SEAT 9G, FLIGHT 1482

When I got to the ground I wanted to stand up and get away from the plane, because I was directly underneath it. I couldn't put any pressure on my left foot. I remember falling and scurrying and falling, and somewhere along the line I got far enough down, [and] a gentleman came along, and I put my arm over his shoulder [and] he carried me out into the field.

FEMALE, AN OFF-DUTY FLIGHT ATTENDANT, SEAT 1D, FLIGHT 1482

We stood out there . . . for the longest time saying, "Where are the ambulances?" They didn't come, and I don't know if [they] couldn't find us or what happened, and I know they said, "Well, time seems like it stands still," but it was a good ten minutes, and I couldn't figure out where they were. I said, "This man needs help. I mean, he's really burnt." And then that eighty-three-year-old woman was out walking around. She was like in a daze, and she said, "You know, my head really hurts. My head's burnt." Her head was burnt. Her hands were burnt. And so I talked to her, and I said, "Just keep putting snow on your hands." I said, "It will help the burn some," and then some guy started yelling, "Bring the people over here for ambulances." I said, "Gary, do you think you can walk over there?" So, we just started walking over and away from the airplane. We got to the edge of the grass, and they still weren't there. The firemen was just telling us, "Come over and stand over here." I said, "This guy needs help. We don't have time to stand," . . . and [Gary] was starting to get dizzy, and I said, "Do you want to sit down?" And he said, "Yes, I would." There was a car sitting there. I think it was a police car. I said, "Why don't you go in [the car]. We'll just sit you in the back. Come sit down." And then all of a sudden it started pouring rain, and everybody just like that left us.

MALE, FLIGHT 1482

. . . A few minutes later I was going to leave [the elderly woman he had helped off the wing] to help somebody. . . . And I saw this one guy standing over there, and I recognized him [from work. I did not know he was on the airplane]. So, I called over to him. I said, "I didn't know you were even on this airplane." He said, "I didn't know you were on the airplane." I said, "My God, how did you get off?" He says, "I . . . jumped off that wing." And I says, "Well, I did too." "Yeah," he says, "I've been over there helping this other guy catch people." I said, "Hell, that was me!"

FEMALE, AN OFF-DUTY FLIGHT ATTENDANT, SEAT 1D, FLIGHT 1482

I didn't know where anybody went. Everybody just like went every which direction. So we got the lady, I think it was the lady in 2D. She was sitting in the back [of the police car]. I got Gary sitting in the back, and the old burnt lady up, the eighty-two-year-old lady was sitting in the front seat. It's pouring rain, and everybody just like left them, and I know it's pouring rain but I got in the car and I said, "I don't know what's going on," but the car was pretty close to the plane. I thought, "I'm getting out of here." I got in the car, and I said, "Well, if nobody else can do it, women can do it. Let's go." So I got in the car and just started driving toward the ambulances away from the airplane. Somebody's got to do something. They're leaving us here, so I thought, "I don't want to be beside this plane," and I don't even know if they knew we're sitting in this car, so I thought, "If I just drive toward where the other police cars are, they'll bring it up, so I did, and I don't know if the [policeman] noticed me. I was moving his car. He came up to me. I said, "I don't know whose car this is. I don't know what's going on but everybody left us and I wanted to bring the car to bring these people closer to the other cars," and he said, "It's my car, but that's all right, just keep them in there."

ELDERLY FEMALE, SEAT UNKNOWN, FLIGHT 1482

The [men who helped me off the wing] said, "We have to walk away [from the aircraft]." It looked like a field. It was snowing and slushy. [There was] some grass, I think. [There was] more than just concrete. And we walked out . . . and we saw some other people. One person they had lying down in that slushy soup, and I thought, "I'm glad I'm on my feet."

FEMALE, AN OFF-DUTY FLIGHT ATTENDANT, SEAT 1D, FLIGHT 1482

The old lady said, "I can't breathe." So I turned the [police car's air-conditioning] vent up full blast. I said, "Well, is that any better for you?" And she just said, "It smells like smoke," and I said, "Well, it's going to," but at least the air was starting to circulate through the car, and then I had kind of got out of the car. I still didn't know where the ambulances were, and I said, "This guy's got to get help. You got to help this guy. You got to put this man into the police car and just take him. Don't wait for the ambulances. This guy's going to go in shock," and I knew he was starting to get worse. I kept trying to talk to them and make all three [wounded in the car] talk to each other. . . . [Then a] guard [came]over and [he] stuck them in a car and . . . that's when I knew

[Gary] was really starting to get bad, because before he could move his legs, and he was all right, and he kept saying, "My chest is really hurting. My chest is hurting," and I think maybe what that was, was from something in the smoke and that was starting to hurt him. . . . I said, "Gary, you'll be all right. They're starting to take you to the hospital now. You've got other people in the car. Just keep talking. . . . Gary, your legs are all right. Just watch your back and get into the car. They're gonna take you. You'll be fine," and we shut the door and they left.

. . . I felt better then because I knew he was going into shock, and I told the guy, "This guy is going to go. You got to get him there." And then I walked over to see if anybody else was still out walking around, and I didn't see anybody else hurt. . . . I stood around a fire truck or ambulance watching, and they were still trying to get a guy on a stretcher coming back off the field. . . . I says, "Is there anybody else left? Is anybody still around? Is everybody all right?" And they're all like, "Yes." It was kind of raining, but then it stopped, and then I was standing there, and I ended up talking to a guy. . . . He got out of a car, and he was on the plane, and I remember his pants were ripped and his boxers were waving outside his pants and we were kind of laughing, because he had cut his butt, and I said, "I know you don't want me to look at it, but you might want somebody to look at it, just to make sure you're all right." So we were laughing. . . .

MALE, FLIGHT 1482

. . . They took us to the terminal and [we] went into like [a] hospitality-type suite [of] rooms, something like that. And I saw a phone on the wall, so I grabbed the phone. You know, you got to help yourself around here, so I did, and I grabbed the phone and I punched the nine, which I thought would get me an outside line, and I did, and I called my wife. My wife answered the phone and she says, "Where are you?" And she says, "Well, do you know that there was an airplane crash out there?" And I says, "Yeah, I know it. Because I was in it." And she says, "Oh, my God."

. . . We were pretty much sitting there. They [the airline personnel] were talking about people. "Do you think you should go to the hospital?" I says, "Well, I don't think I have to go to the hospital after all. I got the burn on my finger. I feel pretty good everywhere else, other than I'm kind of rattled and all that, but probably going to the doctor at the hospital is something I should do for my hand here because you don't know for sure how . . . you don't want an infection. You got to take care of it right." They said, "Yeah." One of the Northwest personnel said, "Well, we'll insist to encourage you. You should go to the hospital for that finger and all that." So I said, "OK, I agree with what you're saying. I'll go to Providence Hospital up near my house. The only thing I need is for somebody to help get me to my car, because I don't have any clothes, and I'm all wet and my feet are wet and that." So, God, they went around and around and back and forth as to how to get me to my car, and this and that and finally somebody says, probably intelligently, "Let's not take him to his car because he shouldn't be driving." I said, "Yeah, that sounds like a good decision." I may think I can drive but I'll probably drive right out of there into a bus or something. So I says, "Well, how do you suggest I get home?" They said, "Well, we'll get you a taxicab. We'll make out this piece of paper and we'll have somebody go out with you and get the taxicab driver and that." I said, "That sounds fine." So, some young girl walked out with me, and that's what we did. I got in the taxicab and went on home.

13. MAUI, HAWAII

April 28, 1988

Aloha Airlines Flight 243

Aloha Airlines flight 243 departed Hilo, Maui, on schedule at 1:25 P.M. en route to Honolulu, Hawaii. There were five crew members—two pilots and three flight attendants—plus an FAA air traffic controller in the cockpit jump seat, and eighty-nine passengers. The flight climbed without incident to a flight level of two four zero (twenty-four thousand feet), with the first officer at the controls, while the captain was attending to non-flying pilot duties. As the flight reached its flight level, both pilots heard a loud "clap" and a "whoosh" sound followed by rushing air behind them. The first officer felt her head jerk back. Debris flew in the cockpit. The door was gone. There was "blue sky" where the first class cabin's ceiling was meant to be. Due to metal fatigue and corrosion, the entire upper fuselage of the aircraft tore off, from the cockpit door to the back of first class, leaving those passengers essentially riding in the open air. The captain started an emergency descent, extended the speed brakes on the wings and dropped airspeed. The first officer tried to declare an emergency over the radio but because of the loud level of noise she did not know whether her declaration was heard. The pilots communicated back and forth with hand signals. The captain was able to make a "normal" landing at Honolulu International Airport. An emergency evacuation was accomplished: Seven of the eighty-nine passengers and one flight attendant suffered serious injuries. One flight attendant died when she was thrown from the aircraft.

FEMALE, SEAT 4A

As best I can recall, we boarded in Hilo on kind of a drizzly day and left with a beautiful takeoff, and we went up [climbed] for about fifteen minutes. . . . The stewardess . . . came to bring our sodas—myself and two young gentlemen who sat beside me. There were no other passengers in first class. . . . [The flight attendant] returned to collect [our cups] and serve any other drinks if we wanted, so she took my glass from me, and I thanked her. . . . I reached up to lock my tray into the table . . . and then there was this thunderous explosion, and I was startled, because there was a great big jolt, and I looked up. My hair was flowing up into the sky and just above me were clouds. I thought we had lost the whole front end of the plane. . . . My arm was outside, and I thought I looked out and saw clouds. . . .

FLIGHT ATTENDANT MICHELE HONDA

Everything was normal, and we had taken off, and we had just completed our beverage service, and I had just gone to the back and sat down and looked up in the rearview mirror and saw my first flight attendant with some purple trash-can plastic bags. So I immediately went to the closet and got my bags to start picking up our beverage supplies from the passengers, and I closed the door and started to go up forward, and then I turned around so my back was to the cockpit and started walking backward collecting cups from the passengers. I assume I was around midsection of the aircraft [when] I heard a loud "pop" and simultaneously I was knocked . . . almost out of my shoes and . . . the next thing I knew, I saw a hand reach out to grab for me, and I grabbed on to the metal retainer bar on the bottom of whatever row I was at.

FLIGHT ATTENDANT AMY LYNN JONES

[I was] just reading through a magazine. The next thing I felt was . . . was just like a gush of wind, and I was still not sure what happened, and [I] looked forward . . . and saw just the hole opening [in the fuselage]. It was not an explosion. Mostly it was the loud noise of wind [and] the sound of the aircraft's engines maybe, but there was no type of explosion noise.

MALE, SEAT 5F

. . . The guy sitting next to me and I were talking, and I remember somebody came by and asked him if he wanted something to drink, and he said, "No thanks," and that is why I recall that they were serving beverages, and that is all I remember. . . . I got carried to the hospital. I still don't remember a thing.

COCKPIT VOICE RECORDER

Sound of screams; sounds of wind. The CVR does not record any of the crew's conversation for the first five minutes after the separation of the ceiling from the fuselage. However, the CVR recorded the crew's transmissions with Maui Control through the cockpit crew's oxygen-mask microphones.

FO: Aloha two forty-three, we're going down. . . . Request lower [altitude]. Center, Aloha two forty-three. Center, Aloha two forty-three. Maui Approach, Aloha two forty-three. Maui Tower, Aloha two forty-three. Maui Tower, Aloha two forty-three. We're inbound for a landing. Maui Tower, Aloha two forty-three.

MAUI: [Flight] calling tower, say again.

FO: Maui Tower, Aloha two forty-three, we're inbound for landing. We're just, ah, west of Makena. . . . Just to the east of Makena, descending out of thirteen [thousand feet], and we have rapid depr—we are unpressurized. Declaring an emergency.

MAUI TOWER: Aloha two forty-three, winds zero four zero at one five. Altimeter two niner niner niner. Just to verify again, you're breaking up. Your call sign is two forty-four, is that correct? Or two forty-three?

Here the crew, having descended to eleven thousand feet, takes off its oxygen masks.

FO: Two forty-three Aloha—forty-three.

TOWER: Two forty-two, the equipment is on the roll. Plan [to approach] straight in [on] runway 2, and I'll keep you advised on any wind change.

FO: Aloha two forty-three... [To CAPT] Do you want me to call for anything else?

CAPT: Nope.

TOWER: Aloha two forty-four on the emergency?

FO: Aloha two forty-three.

TOWER: Ah, two forty-six.

FO: Aloha two forty-three.

TOWER: Aloha two forty-three, say your position.

FO: We're just, ah, to the east of Makena [at a] point descending out of eleven thousand. Request clearance into Maui for landing. Request the [emergency] equipment.

TOWER: Okay, the equipment is on the field . . . is on the way. Squawk zero three four three. Aloha two forty-three, can you come up on [frequency] one one niner point five?

FO: Two forty-three, can you hear us on one nineteen five two? Maui Tower, two forty-three. It looks like we've lost a door. We have a hole in this, ah, left side of the aircraft. [To Captain] Want the [landing] gear?

CAPT: No.

FO: Want the landing gear?

CAPT: No.

FO: Do you want it [the landing gear] down?

CAPT: Flaps fifteen [for] landing.

FO: Okay.

CAPT: Here we go. We've picked up some of your airplane business right there. I think that they can hear you. They can't hear me. Ah, tell him, ah, we'll need assistance to evacuate this airplane.

FO: Right.

CAPT: We really can't communicate with the flight attendants, but we'll need [fire] trucks, and we'll need, ah, air stairs from Aloha.

FO: All right. [To Tower] Maui Tower, two forty-three, can you hear me on Tower?

TOWER: Aloha two forty-three, I hear you loud and clear. Go ahead.

FO: Ah, we're gonna need assistance. We cannot communicate with the flight attendants. Ah, we'll need assistance for the passengers when we land.

TOWER: Okay, I understand you're gonna need an ambulance. Is that correct?
FO: Affirmative.
CAPT [to FO]: It feels like manual reversion.
FO: What?
CAPT: Flight controls feel like manual reversion. . . .
FO: Can we maintain altitude okay?
CAPT: Ah . . .
FO: Can we maintain altitude okay?
CAPT: Let's try flying . . . let's try flying with the gear down here.
FO: All right. You got it.

Sound of gear being lowered.

TOWER: Aloha two forty-three, can you give me your souls on board and your fuel on board?
CAPT [to FO]: Do you have a passenger count for the tower?
FO: We, ah—eighty-five, eighty-six, plus five crew members.
TOWER: How many do you think are injured?
FO: We have no idea. We cannot communicate with our flight attendant.

FLIGHT ATTENDANT MICHELE HONDA

I hung on to the [seat] bracket, and there was someone holding on to me as well, and during this time, I think I heard not another popping sound so much as what I presume [was] the fuselage [being torn off the top of the aircraft]. I did not see anything. I was on . . . the floor, and we had begun our descent, and I was just trying to keep my breathing so I would not pass out, and in the meantime I was trying to look up for an oxygen mask, but that was not as important as trying to keep my breathing.

There was quite a bit of debris. [I] could see some little bit of vapors, and I remained in that position until I felt the plane leveling off. And then at that time, I pulled myself up and I looked but the wind was really strong and pushing me back, and I saw the south side of Maui, and . . . I saw ocean and . . . everything was exposed. . . .

I noticed that my third flight attendant was incapacitated and that there was blood. There was debris over her, and other passengers were holding her down at that point, so when I could not get up to her initially, I started to go back down, because the wind was so strong and it removed debris as I went down, caught my breath again and proceeded to go back up and removed more debris, and I saw a hand going like this to "keep down," so I started to tell passengers to keep down, [keep their] heads down. People were starting to put on life vests, and I was helping people crawling up and down [the aisle] holding on to the bars [of the seats]. Then I went toward the back and tried to get to the cockpit interphone but nothing was happening. So I proceeded back down [the aisle], removed debris. Again, I was hoping to keep down and try to get to the front, and I was struggling with the life raft that was in the middle of the aisle, and hands came out and they pushed [the life raft] out of the way for me so I could try to get to [the incapacitated flight attendant] Jane.

. . . It was really intense, and [Jane] was hanging on. I did not realize at the time that she was out—not conscious. And I tried to pull her back and could not, but there were people holding her down, so I went back down [the aisle] and helped with life vests.

FLIGHT ATTENDANT JANE HOLUMI SANTOTOMITA

. . . The next thing I remember I was on the floor in the aisle, and I saw feet and passengers on the side saying, "Grab her," and so these two men grabbed my hand and held on to my hands, and we were squeezing each other's hands.

I opened my eyes, and then I looked up, and then I saw the blue sky, and I was really scared, and I saw all the [damage] and all the damaged seats and floor . . . with all the wires sticking out. The thing is, I saw the blue sky and the torn roof, and I was really scared. I said I did not want to die. And I closed my eyes. And then they said it took about thirteen minutes to land the plane, but to me it took about two minutes.

FLIGHT ATTENDANT AMY LYNN JONES

. . . I looked out my window, and I saw water, and I did not see land. I don't know how far into the flight we were. I am guessing about ten or fifteen minutes. I assumed that Maui would be [coming up] shortly, but I did not know how far out we were, so I was getting people to get their life vests on.

FEMALE, SEAT 4A

I had no feeling in my arm and pulled it back in, looked down, and realized I was going [out of the plane] with [my arm] because my seat had tilted into what looked like on a [single] rail, and my seat was stuck in that rail, but it was jostled from where it should have been, and the gentleman beside me was putting on his life vest, so I tried to reach for mine. But because I was sitting at an angle, I could not reach for [my life vest], and I told [the man beside me] I had no [life vest], so [the gentlemen beside me] both reached over and hung on to me. And then I put my arms around the fellow next to me, because he let go of me and tried to find me a life vest under my seat, and he said [the life vest] was gone. So he leaned forward while I clamped myself into his vest, and the fellow on the aisle reached over behind him and just clung to me for dear life. And by their very reaction [alone] I did not go out that window. But immediately after the explosion when I realized that there was nothing around us, immediately there were little red . . . objects kind of moving across and I didn't want to get [them] on me. There was no real smell of burning. There was just a whole bunch of cables on the third fellow, and a cable had come down, but the rest of the ceiling was gone.

I then realized that we were up that high [altitude and] I tried to breathe shallowly so that I would not be without breath, as I held on, and then when I started to hold on to the gentlemen [beside me] I realized I was covered in blood, because we were getting all bloody, and it was mine.

FLIGHT ATTENDANT MICHELE HONDA

Things get real blurry right here. I don't know how many times I went up and down [the aisle]. I do recall individual passengers needing help. One lady [was] asking me to help her son with his life vest. . . . Another man was struggling with his life vest and I tried to help him put it over [his head] but he said, "Wait a minute, can you remove this?" and part of the side of the aircraft was stapled into his face, and I did not realize it, and when I tried to pull it away, his face came with it. I said, "I can't remove it right now," and at that time I think his wife said, "Honey, I will help you." So I continued down the [aisle]. Another passenger asked me, "Do we have a pilot?" I said, "I don't know, I don't know." He said, "Well, do we have a pilot?" And I said, "I don't know."

COCKPIT VOICE RECORDER

TOWER: Okay. We'll have an ambulance on the way.

FO: There's a possibility that, ah, we won't have a nose gear.

CAPT [to FO]: You tell 'em that we got such problems, but we are going to land anyway even without the nose gear. But they should be aware of . . . we don't have a safe nose-gear-down indication.

TOWER: Aloha two forty-three, wind zero five. The [emergency] equipment is in place.

FO: Okay, be advised, we have no nose gear. We are landing without nose gear.

TOWER: Okay. If you need any other assistance, advise. . . .

FO: We'll need all the [emergency] equipment you've got. [To the captain] Is it easier to control with the flaps up?

CAPT: Yeah. Put 'em at five. Can you give me a Vee speed for a flaps-five landing?

The FO can't find the reference in the manuals.

FO: Do you want the flaps down as we land?

CAPT: Yeah, after we touch down.

FO: Okay.

The captain and the FO discuss the speeds for landing.

TOWER: Aloha two forty-three, just for your information, the gear appears down, gear appears down.

FO [to captain]: Want me to go flaps fort . . . . ?

CAPT: No.

FO: Okay.

Sound of touchdown on runway.

FO: Thrust reverser.
CAPT: Okay, okay, shut it down.
FO: Shut it down.
CAPT: Now, left engine.
FO: Flaps.
TOWER: Aloha two forty-three, just slow her down where you are. Everything [is] fine. The gear did . . . The fire trucks are on the way.
CAPT: Okay. [To FO] Okay, start the call for the emergency evacuation.

FLIGHT ATTENDANT MICHELE HONDA

. . . After we landed, I tugged on Jane. I said, "Come on, let's go!" Her leg was wrapped around one of the cables, but after another passenger undid the cable around her leg I pulled her and she started to run after me, which was real good, because for a second I thought maybe she could not move. But she did, and Amy [another flight attendant] had jumped out of her seat at that time, as we were running toward the back [of the aircraft], and I said, "Take care of the air stair." I deployed the aft slide. It went off but the wind blew it to the side [of the aircraft], so I called the man in the [firefighter's] outfit, and he straightened it out, and I sent Jane down first because she was incapacitated, and I had told the passengers after we ran to the back to wait, because I did not want them to come out because nothing was out for them to [exit, evacuate on] yet. [The chute] was still inflating, but they were already jumping out of their seats, and as soon as [the chute] was available, I called them, and they came rushing, and we used both exits and I think the rest of it is kind of history.

FEMALE, SEAT 4A

We hung together, and I don't know how long it took us, but we were down on the ground, and [we made] the most beautiful landing. I have to say, I thought we had no pilot, no front of the plane, no nothing. . . . In that long period when I was holding on to those gentlemen I could hear the plane's engines still running, and I could not believe it, but in any case, we did land [and] it was just beautiful. There were no bumps, no nothing. He came in so beautifully. The whole time that we were up there [in the air] I could hear this rushing of winds, like a wind tunnel. I have never experienced anything like this. I have no words for it.

I say the arm of my chair was ripped off. I am sure [of that], because I was there. I felt myself kind of sliding off [the chair] and had to hang on for dear life, but when I looked down all I could see was the [chair] rail that the chairs are supposed to be inside of. . . . That is all that I could see there. And, of course . . . way down there all of that beautiful mass of ocean. I am very appreciative of the fact that [the gentleman who held on to me preventing me from flying out of the aircraft] said, "You know, we thought you were going over the side, and we decided that we had better hang on, because if you have to go, we are all going to go together." And so, it was because of their actions that I am still here.

14. SIOUX CITY, IOWA

July 19, 1989

United Airlines Flight 232

United Airlines flight 232, a DC-10-10 wide-bodied aircraft en route from Denver to Chicago, while flying at a flight level of thirty-seven thousand feet at 3:16 P.M. and about two hundred miles west of Dubuque, Iowa, experienced an explosion in its number-two engine. The separation of the engine, the fragmentation of its fan blades and other parts, and the explosive force of the fan rotor parts severed the aircraft's hydraulic lines, which control the wing flaps, elevators, ailerons, and rudder. In short, all usual control mechanisms were destroyed in the explosion. For the next forty minutes the crew of flight 232 maneuvered the aircraft with the throttles of the two remaining [wing] engines. Flight 232 diverted to Sioux Gateway Airport, where it touched down on the threshold and to the left of the center line on runway 22. The first contact with the ground was made by the right wingtip. The right main landing gear broke the concrete beyond the first impact point. The main fuselage section stopped upside down in a cornfield between runway 17 and taxiway L. Both wings were still attached, though most of the right wing had broken off. The tail section had broken off at the first impact and continued to slide and tumble down the runway until coming to a stop on Taxiway L. The main fuselage section and the tail section came to a stop 3,650 feet from the point of first impact. Of the 296 passengers aboard flight 232, 183 survived the accident, including seven flight attendants and three cockpit crew members. Fatalities included one flight attendant and 110 passengers who died within thirty days of the accident; one passenger died thirty-one days after the accident.

The passengers on flight 232 had little indication of their peril. The captain of the flight, Al Haynes, a fifty-eight-year-old veteran of commercial aviation who was later credited with saving the aircraft, kept the passengers informed to the extent he thought necessary to reassure them. Meanwhile, the drama in the cockpit went unseen and unheard by the passengers.

In the aircraft cabin, the first flight attendant heard the engine explode while she was picking up trash, and she immediately sat down on the floor until the aircraft seemed "normal" again. The captain made an announcement to the passengers that the airplane had lost an engine. The first flight attendant stood up and reassured a passenger sitting in seat 23A who was clutching a baby in her arms and a woman in 23E who looked terribly frightened. The captain then called the first flight attendant on the interphone and asked her to come to the cockpit. There he told her the situation. He ordered her to prepare the cabin for evacuation. She did not ask any questions because, as she later stated, "The cockpit crew was working very hard." She did not gather the other flight attendants together, because she did not want to alarm passengers. She informed each of them individually, telling them to secure meal-service equipment as quickly as possible. They picked up the trays.

The first flight attendant was aware of an off-duty United pilot sitting in first class. He was wearing civilian clothes. A passenger pointed out the window to show her the right horizontal stabilizer, which was visibly damaged. The first flight attendant returned to the cockpit and described what she just saw. The captain told her that the landing would be "quick and dirty."

Her hands were shaking as she put away meal trays. She postponed an announcement of the evacuation plans until she was "sure that [she] could stand there and do it in the manner that [she was] supposed to." She read the quick-preparation speech to the passengers over the PA system, adding that the infants on the aircraft should be placed on the floor.

Meanwhile, in the cockpit, the crew was struggling to maneuver the aircraft into position for a landing.

COCKPIT VOICE RECORDER

CAPT [to Sioux City Approach Control]: Ah, we're controlling the turns by power. I don't think we can turn right. I think we can only make left turns. We're starting a little bit of a left turn right now. Maybe we can only turn right. We can't turn left.

APPROACH: United 232 heavy, ah, understand you can only make right turns.

CAPT: That's affirmative.

APPROACH: United 232 heavy, roger. Your present track puts you about eight miles north of the airport, sir. And, ah, the only way we can get you around [to runway 31] is a slight left turn with differential power or if you go and jockey it over.

CAPT: Okay, we're in a right turn now. It's about the only way we can go. We'll be able to make very slight turns on final, but right now just . . . we're gonna make right turns to whatever heading you want.

MALE, AGE FORTY-TWO, SEAT 25G

Loud explosion. Captain told us we had lost number-two engine and would be a little late getting into Chicago. Later told us we damaged tail and would make an emergency landing in Sioux City. "I'm not going to kid you," [the captain] said. "This is going to be a very rough landing."

MALE, AGE SIXTY-THREE, SEAT 28H

The flight was smooth—sky clear. I had just finished lunch, my tray on the table. My seat belt was fastened. . . . There was a sharp report, not the boom associated with an explosion. At the same time, I noticed a flash outside the window. The aircraft vibrated for a split second and immediately banked to the right and nosed down slightly. Within seconds, the aircraft was stabilized, and we were informed that the number-two engine had exploded. We were also informed that our arrival at O'Hare will be a few minutes late. The plane then began to drop the right wing. At times, the wing would drop quite steeply. By this time, we were down in a sky full of tall white clouds. I felt that the crew were avoiding entering these clouds. The cockpit informed us that they will inform us when we are ten minutes from touchdown, and then they will give us a four-minute warning and a four-second warning. The cockpit also informed us that after the four-second warning, we will be told, "Brace, brace, brace."

MALE, AGE FORTY-TWO, SEAT 11E

About forty minutes from Chicago, I was sitting with my seat belt loosely on, since I had just let the man sitting next to me get out to go to the rest room, when there was a very loud bang and a jolt that knocked the attendants and everyone in the aisles to the floor. We plunged from whatever altitude we were cruising [at] to about cloud level very quickly. Everyone began gathering up briefcases and stowing them overhead, as I let the gentleman back to his seat. I sat back down and strapped my seat belt very tightly. The pilot came on the PA and announced that we had lost the number-two engine and would be late to Chicago. Several minutes later, we began to hear left and right as each wing engine alternately whined loudly. One of the officers of the crew kept running back and forth to the rear of the plane several times. About ten minutes before the crash, the captain got back on the PA and apologized for taking so long and said that the number-two engine had caused damage to the tail and we were making an emergency landing at Sioux City, Iowa, and that he wasn't going to kid us that this could be a rough landing, maybe a *very* rough landing and that the flight attendants would describe emergency landing and exiting procedures and that at four minutes to touchdown the crew would yell "Brace" three times, which they did.

My wife saw the look on my face and said everything would be all right in order to prevent me from scaring my son.

FEMALE, AGE TWENTY-FIVE, SEAT 14J

A stewardess had given me a blanket to cover my son with, and a female had told us on the PA to place infants on the ground at our feet and hold them there. My son was asleep, and I put him in front of my feet with his head toward the right side of the cabin, away from the aisle. I believe he was on his back. About two minutes prior to the crash, the pilot repeated "brace" three times, and I heard a female voice repeating the words.

COCKPIT VOICE RECORDER

CAPT: How do we get the [landing] gear down?

OFF-DUTY PILOT HELPING OUT IN THE COCKPIT: Well, they can free-fall. The only thing is, we alternate the gear. We got the [landing gear] doors down?

CAPT: Yup.

FO: We're gonna have trouble stopping, too.

CAPT: Oh, yeah. We don't have any brakes.

FO: No brakes?

CAPT: Well, we have some brakes [but not much].

OFF-DUTY: [Braking will be a] one-shot deal. Just mash it, mash it once. That's all you got. [To captain] I'm gonna turn ya. [I'm gonna] give you a left turn back to the airport. Is that okay?

CAPT: I got it. [To APPROACH]. Okay, United 232, we're startin' to turn back to the airport. Since we have no hydraulics, braking [is] gonna really be a problem. [I] would suggest the [emergency] equipment be toward the far end of the runway. I think under the circumstances, regardless of the condition of the airplane when we stop, we're going to evacuate. So you might notify the ground crew [pretty much] that we're gonna do that.

Landing is minutes away.

OFF-DUTY: Let's start down. We have to ease it down.

CAPT [to APPROACH]: We're startin' down a little bit now. We got a little better control of the elevator.

APPROACH: United 232 heavy, roger. The airport's currently at your one o'clock position, one zero [ten] miles.

CAPT: Ease it down, ease it down.

Sound of groan. Sound of exhalation.

OFF-DUTY: I got the runway, if you don't. . . .

CAPT: I don't. . . . Come back [on the yoke]. Come back.

OFF-DUTY: It's off to the right over there.

FO: Right there. Let's see if we can hold [a descent rate of] five hundred feet a minute.

APPROACH: United 232 heavy, if you can't make the airport, sir, there is an interstate [highway] that runs north to south, to the east side of the airport. It's a four-lane interstate.

OFF-DUTY: See? We got [the] tower [in sight] right here at our one o'clock low. . . .

CAPT [to APPROACH]: We're just passing it [the highway] right now. We're gonna try for the airport. [To OFF-DUTY] Is that the runway right there? [To APPROACH] We have the runway in sight. We have the runway in sight. We have the runway in sight. We'll be with you shortly. Thanks a lot for your help.

OFF-DUTY: Bring it down. . . . Ease 'er down.

FO: Oh, baby.

OFF-DUTY: Ease her down.

CAPT: Tell [the passengers on the PA] that we're just two minutes from landing.

APPROACH: United 232 heavy, the wind's currently three six zero at one one three sixty at eleven. You're cleared to land on any runway. . . .

CAPT: [Laughs] Roger. [Laughs] You want to be particular and make it a runway, huh?

FLIGHT ENGINEER [on PA]: Two minutes.

OFF-DUTY: What's the wind?

CAPT [to APPROACH]: Say the wind one more time.

APPROACH: Wind's zero one zero at one one. . . .

FO: Yeah, we want to go down.

OFF-DUTY: Yeah, I can see the runway but I got to keep control on ya.

FO: Pull it off a little.

CAPT: See if you can get us a left turn.

FO: Left turn just a hair, Al.

CAPT [to APPROACH]: Okay, we're all three talking at once. Say [the wind] again one more time.

APPROACH: Zero one zero at one one, and there is a runway that's close, sir, that could probably work to the south. It runs northeast to southwest.

CAPT: We're pretty well lined up on this one here.

OFF-DUTY: I'll pull the spoilers [speed brakes] on the touch [down].

CAPT: Get the brakes on with me.

APPROACH: United 232 heavy, roger, sir. That's a closed runway, sir, that'll work, sir. We're gettin' the equipment off the runway. They'll line up for that one.

CAPT: How long is it?
APPROACH: Sixty-six hundred feet, six thousand six hundred. Equipment's comin' off.
CAPT: Pull the power back. That's right. Pull the left one [throttle] back.
FO: Pull the left one back.
APPROACH: At the end of the runway it's just wide-open field.
COCKPIT VOICE: Left throttle, left, left, left, left. . . .
COCKPIT VOICE: God. . . .

Sound of impact.

MALE, AGE FORTY-TWO, SEAT 25G

Jarring impact.

Lost glasses on impact. Plane appeared to accelerate after impact and began to roll. When plane stopped, lights went out, and I was hanging from the ceiling. Plane upside down. Something began to fall on top of me which I deflected with left arm. I thought the cabin was collapsing, and we would all be crushed.

MALE, AGE SIXTY-THREE, SEAT 28H

After the order to "brace," I saw right wing dig into the dirt with fire at the tip of the wing. Moments later, the wheels hit the ground hard. We were informed earlier that the landing will be hard. I felt being airborne again and then we hit hard a second time. I must have been knocked out or blacked out. There was a lot of noise on impact.

FEMALE, AGE TWENTY-ONE, SEAT 37D

The whole accident was so unusual. I remember being jolted around like in a [clothes] dryer. The impact was so intense. I have never felt such pressure before in my life. I saw sparks and glass flying everywhere. I was so scared for all the lives of the passengers. I knew immediately that we were in serious trouble. For some reason, my hand was thrown back. I was holding my ankles tightly.

FEMALE, AGE FORTY-SIX, SEAT 33D

I remember the severity of the impact, the decompression, and the subsequent rush of air and debris out of the plane, and then the large amount of dirt flying into the plane. I was tossed about and had a difficult time continuing to hold on to the seat in front of me.

MALE, AGE FORTY-TWO, SEAT 11E

The first impact was so violent that my chest and ribs went crashing down on my knees, knocking the wind out of me and sending sharp pains through my ribs. It knocked me upright a little, and I saw the flames shoot over our heads at about eleven o'clock. I looked back and to the left across the aisle to my wife and son, when some kind of pane came slashing down like a guillotine next to her seat, blocking them from my sight and blocking them from the aisle.

FEMALE, AGE TWENTY-FIVE, SEAT 14J

The plane hit the ground with a large impact twice and began rolling to the right. I heard metal scraping the ground and I remember dirt, plastic flying all around me. My son flew up in the air, and I managed to grab ahold of him around the waist. He struck his head several times before the plane came to a stop and several times I had to pull him back into my arms as he slid out of my grip.

MALE, AGE FORTY-TWO, SEAT 25G

I released seat belt to escape being crushed and fell to the floor-ceiling. A man's voice said, "There's fire. We can't go this way," which was forward from my seat. There was no movement in the other direction. Even without [my] glasses, I could see fire outside the window.

Smoke began to accumulate. I could not understand why there was no firefighting equipment. I saw no way to escape. I had no sense of others [around me]. I thought I would burn to death. I crouched to the floor to breathe. When I looked up, I saw a light haze to the rear of the plane where before it was dark. I walked to approximately row 30 and stepped into a cornfield.

MALE, AGE SIXTY-THREE, SEAT 28H

When I was unbuckling myself, all passengers to my left were gone. The two passengers in front of me were gone. The passenger to my right was upside down, legs in the air. I helped her up. I think I may have blacked out immediately and was unaware of passengers around [me].

The passenger whom I helped get up began walking toward the rear of the cabin, which was pitch-black with smoke. I attempted to get out by walking forward where daylight and less smoke [were] visible. A panel hanging from the ceiling was dangling in front of me. In attempting to push the panel aside, I received minor burns to my fingers. The fire outside was hot. The panel actually protected me from the heat. On several occasions, I had to get down to the floor level to get fresh air.

After several attempts, a male voice asked if he might pull himself up by taking hold of my leg. I said OK. At this point, I felt something cold hitting me. It turned out to be from [foam? water?] being directed at the fire. At this point, the other passengers and I walked out at the point where the fuselage broke away from the front section. While walking out, my shoes got tangled in the twisted metal and came off.

FEMALE, AGE TWENTY-ONE, SEAT 37D

I had trouble getting out of my seat belt. My hands just could not—I guess that they were not strong enough. The glass and debris was terrible. I am uncertain as to what hit me. But I knew that [I] was injured.

. . . Somehow a big hole was formed. I saw that my nephew was out and then I jumped. I had lost my shoes, so my feet were hurt severely when I jumped into the debris left from the hole. I wish I would have had my tennis shoes on instead of sandals. I wish I could have helped more people.

FEMALE, AGE FORTY-SIX, SEAT 33D

When the motion stopped and I opened my eyes, I saw the ground. My only thought was to get out as fast as I could, so I unbuckled my seat belt and fell/slid about five or six feet to the ground. There was a large amount of metal debris that I jumped over and then I ran for my life. I saw no flames but a vast amount of dense black smoke to my left as I ran toward the airport and the rescue vehicles. One of those rescue workers asked if I'd been on the plane. Only one other passenger was visible to me at this point, and when I said "Yes," he had me lie down on the ground. Soon more survivors began to be rescued.

MALE, AGE FIFTY-SEVEN, SEAT 10H

After coming to a stop upside down, a rolling wall of flame came toward us out of the first-class area. It got within five feet of us and went out! My wife released [her seat belt] and fell heavily to her neck and shoulders. I released [my belt] and held on, so I landed on my knees, right-side up. I smelled fresh air and corn to the rear of the plane. I told my wife to stay right behind me, and we crawled back to an opening that had been torn in the plane wall. We stepped into a soft damp cornfield. I put my wife in front of me, and we started running through the corn rows. We hit a clear spot and heard a loud explosion behind us. [We waited] with others at the edge of the airport until [we were] picked up by [the] Air National Guard and [we] were taken to a mess hall.

MALE, AGE FORTY-TWO, SEAT 11E

When we stopped, the aircraft was upside down, but I could only think of trying to free my family, so when I released my seat belt, I was shocked to have the sensation of falling to the *ceiling*. I tumbled and was disoriented and began to look for my wife and son among the cables and debris on the ground level instead of overhead where their seats were. The Oriental gentleman and his child who were sitting behind my wife couldn't get past me and directed me to the hole in the craft. I exited through the hole in the cabin and turned around and looked up at the smoke pouring out of the top right corner of the overturned craft and cried out twice, "My wife and my son are in there."

After quite a few passengers exited, I saw my son and my wife come out, and we ran down the rows of the cornfield, afraid that the wreckage would blow up. We came to a clearing where many [passengers] began congregating, and as I attempted to tie my shoe I had a sharp pain in my ribs. I saw a helicopter fly over, and later on saw the water pumping over the wreckage.

FEMALE, AGE TWENTY-FIVE, SEAT 14J

When the plane stopped, it was totally black inside, and we were upside down. I released my seat belt and, still holding my son, who was twenty-three months old and [weighed] twenty-two pounds, we fell to the floor-ceiling, where he again bumped his head and I my shoulder. My seat partner assisted us by pushing us to an opening into the cornfield. We walked barefoot over plastic, luggage, and were pushed and stepped on until we were outside. At that time, I had to walk through cables hanging from the aircraft. Once we reached a dirt road between cornfields, I looked back to see smoke billowing from the crash site.

MALE, AGE SIXTY-THREE, SEAT 28H

Once out of the aircraft, I notice a man, woman, and child sitting on the ground. A young man approximately twenty-five [years old] was pacing back and forth at one time, making an attempt to go toward the fuselage.

Approximately fifty yards from the fuselage a man approximately forty years old was lying facedown, breathing, with all of his clothes excepting his shorts burned or blown off [his body]. About ten yards from him another male about the same age was sitting, bleeding from the face. Approximately ten percent of his clothes were burned off.

About two or three minutes later, a National Guard man approached me and asked me to sit down. He then went toward the fuselage but shortly turned around and asked me to walk with him to an assembly area. During this walk, I noticed debris as far as I could see. I also noticed a landing-wheel assembly.

FEMALE, AGE FORTY-ONE, SEAT 15H

It was so terrifying, so amazing to be alive and virtually unhurt physically, everyone running through the corn and once at a road calling for others. [A] stewardess kept us [about thirty survivors] together and said [for us to] wait for others, realizing the numbness of my body, every feeling, everyone walking slowly up and down the road hugging everyone, tears rolling down faces, so glad to be alive, so concerned and afraid for others, seeing such destruction, smoke and fire, hoping so much for more survivors. Oh, we prayed for others and thanked God for our survival. I'll never be able to forget the terror of this crash. Thank God for the skill of our pilot!

15. LITTLE ROCK, ARKANSAS

June 1, 1999

American Airlines Flight 1420

American Airlines flight 1420, a McDonnell Douglas MD-82 on a regularly scheduled passenger flight from Dallas, Texas, overran the end of the runway, went down an embankment, and collided with approach-light structures after landing at Little Rock airport. Thunderstorms and heavy rain were reported in the area at the time of the accident. There were eleven fatalities, including the aircraft captain, and 145 passengers and crew aboard the flight suffered numerous injuries.

COCKPIT VOICE RECORDER

CAPT: I hate droning around visual at night in weather without having some clue where I am.
FO: Yeah, but the longer we go out here the . . .
CAPT: Yeah, I know.

Sound of stabilizer-in-motion horn.

FO: See how we're going right into this crap?
CAPT: Right.
FO [to Little Rock Approach]: American fourteen twenty, I know you're doing your best, sir. We're getting pretty close to this storm. We'll keep it tight if we have to.

Nearly two minutes go by.

CAPT: Aw, we're going right into this.

APPROACH CONTROL: American fourteen twenty, right now we have, uh, heavy rain on the airport. The current weather on the ATIS [Advanced Traveler Information Systems] is not correct. I don't have new weather for ya.

Another minute goes by.

CAPT: Landing gear down. . . . And lights, please.

APPROACH CONTROL: Wind-shear alert, center field wind, three five zero at three two, gusts four five. North boundary wind three one zero at two niner. Northeast boundary wind three two zero at three two.

FO: Flaps twenty-eight?

CAPT: Add twenty.

FO: Right.

CAPT: Add twenty knots.

FO: Okay.

Seventeen seconds go by.

APPROACH: American fourteen twenty, roger. Runway 4 Right. Cleared to land. . . .

Approximately a minute goes by.

CAPT: This is, this is a can of worms.

APPROACH: Wind is three three zero at two eight.

FO: There's the runway off to your right. Got it?

CAPT: No. . . . I got the right runway in sight.

FO: You're right on course. Stay where you're at.

CAPT: I got it, I got it.

Twenty seconds go by.

FO: We're way off.

CAPT: I can't see it.

FO: Got it?

CAPT: Yeah, I got it.

FO: Hundred feet.

The first officer counts down the feet of altitude to ten, when there are sounds of thuds similar to an aircraft touching down on a runway.

FO: We're down. . . . We're sliding.

Unintelligible sounds, then sounds of several impacts.

MALE, AGE SIXTY-THREE, SEAT 7E

A few seconds after touchdown, [when] the thrust reversers had not kicked in and the spoilers [devices on the wings that baffle lift] had not deployed, I knew we were in trouble.

I am a retired Air Force [officer and pilot] and have approximately three thousand flying hours on military aircraft. I also sensed that touchdown speed was quite high. I suspect we were approximately halfway down the runway when I told my wife, "Put your feet up and brace yourself. We are not going to make it."

I felt the plane go off the side of the runway, back onto the runway, then off again on the right side. Shortly after, I saw all the ceiling panels rippling as if a shock wave was passing over them. Then, suddenly, they began to fall. The lights went out. During this time the plane was shaking violently, and I sensed the plane was turning around, even thought I could not tell how far around we turned. My thoughts were, "We're going to flip."

I do not remember if the shaking was before sensing a turning sensation or after. I do suspect that the ceiling panels rippling was at the time the left side of the first class section was being sheared off. I don't know why I did not see the first class section being torn away.

The next thing I sensed, we were completely stopped, and I told my wife, "Get out of here."

I pushed the overhead luggage rack and luggage off of me, pushed my wife out ahead of me, got to my feet, untangled myself from debris—stuff hanging from the ceiling—got in the aisle, picked a man off the floor, and pushed him ahead of me. Then I thought, "Ann [his wife] went out; hope the plane wasn't too high off the ground." I could see, from lightening flashes, the hole where first class . . . sheared off. As I got to the edge of the opening, I saw my wife—from lightning flashes—had fallen when she went out the side opening. I jumped out, helped her up. We were in shin-high grass and were headed directly toward the river. Of course we did not know how close to the river we were. We waded through a knee-deep or deeper drainage ditch to a high spot just a few feet, maybe fifty or seventy-five feet, from the drainage ditch. I kept telling my wife, "Don't lose your shoes," knowing that we were in a very rough area. Everyone huddled in an area, trying to keep warm and trying to help one another.

. . . I got hit very hard on the head, had several small cuts [but] nothing major. I suspect my chin got pushed down into my chest, because I was really hurting in my upper chest and my head. I am still having headaches and my neck gets very tired quite often. . . . My wife says that I was probably in shock, which I agree with. I could not get warm and would shake extremely hard from my head to my feet. One of the passengers pulled off her sweater and put it around me.

FEMALE, AGE THIRTEEN, SEAT 16F

The entire experience was absolutely horrible. I could not believe the lack of help from the other passengers. There were no emergency exits near me. The fire was right behind me, and I got out through a hole in the plane. Adults were yelling at me when I was screaming to get out because the fire was basically lapping at my heels. I strongly think that there should be more emergency exits in the middle of the plane. If that crack had not been there, I would have had to walk through the flames. Also, I think there should be more detailed emergency instructions about fire, because there are none on the [safety instructions] sheet. . . . I am extremely disappointed.

FEMALE, AGE FIFTY-EIGHT, SEAT16D

I knew we were in big trouble when the plane came in very fast and steep . . . and couldn't be slowed down. I'm not a pilot or trained in this, but it felt and sounded like the pilot tried to reverse the engine as many as three times without slowing the plane. I can only imagine his horror at realizing we were hurtling toward a deep and swollen river at one hundred miles an hour. I believe that [the] pilot chose to hit the metal light stands. He saved many lives with that action.

FEMALE, AGE SIXTY-TWO, SEAT 29F

As the plane circled for a landing, we went in and out of a couple of light rain storms. As the plane came down for the landing, we entered a very heavy rain, just before touchdown. . . . I was sure we were going to take the plane up out of the storm. But we landed going fast and did not seem to slow down; fishtailed, apparently were hydroplaning; then [I] heard a thump, bump, bang, several times; then [we] seemed to be off the runway. Then immediately we abruptly stopped. My glasses flew off. Several people behind me headed for the rear exit. The attendant and I, and I don't know who else, tried to open the crumpled exit door. Since smoke and fire were coming our way fast, [we] ran up front. People in rows ahead were saying there was a way out up ahead through the smoke. At the same time, the attendant determined that we could exit at the opening at the top of the crumpled exit door, and only seconds had passed [before] we were evacuating from that opening. Others were calling . . . for us to gather [together]. We held on to each other and ran from the plane to the edge of the water, where we huddled for a long time in the wind, rain, hail and lightning.

MALE, SEAT 10D

Arriving in the Little Rock area [the] captain indicated [that there was a] "light-show" to the left [of the aircraft]. The [lightning] shifted to the right and back a couple of times during the landing prep[aration], indicating a circling maneuver. I grew a bit more concerned and buttoned my wallet in my back pocket. On the final approach, the air got heavy, and I buckled in tight.

The captain landed the plane, and we slid down the runway with lots of bumps, bangs, and [a] shaking sound. It lasted about thirty seconds. During that time, I was planning an escape. I had counted rows [to the nearest exit] . . . prior to arriving in the Little Rock area, and I [had] toured the rear of the aircraft. I looked behind once during the crash and saw the overhead storage opening. The compartments [unhinged] back to front of the aircraft. The luggage [fell out]. I recall several hits to the head and upper body and remember the hit that tore off my glasses. The [engine reverse] thrusters to slow the aircraft came on twice with no effect on me [did not throw him forward against his seat belt]. [The thrusters] were on for three to five seconds and then [were] cut off. The plane leaned to the left about halfway through the landing, and when [the plane finally stopped] I had to adjust my position to release myself from the seat [belt]. The plane was totally quiet. I heard no sound. I saw a lady in about row six get up and disappear to the left. Another lady [in the same row or back one row] also disappeared to the left. I was struggling with the seat belt during that time. I quickly followed them from my seat, up to the opening in the fuselage. . . . As I turned toward the hole, I noted a lady sitting in her seat staring straight ahead. I released her belt and touched her wrist. She got up and beat me to the hole. At the hole, I noticed the grass blown flat and hesitated because I could not recognize the surface without my glasses, but [I] stepped out without knowing what might be underfoot. . . . My exit took about ten seconds from the final jolt and the stopping sensation. . . . I'd say that it was forty or fifty seconds [from the time] the plane stopped until the fireball appeared. It was very windy and rainy with some ice pellets. . . . During the thirty or forty seconds it took for the plane to stop, I had the feeling that there was a lot of activity going on in the cockpit. Just a feeling that the captain and the first officer were trying everything they could and that if they had not been so concerned, many more of us would have died.

FEMALE, AGE 48, SEAT 17D

If the plane hadn't cracked [or] split, I would have died. Fire erupted behind my seat, so the six exits in the rear of the plane were inaccessible. There was no way to reach the two exits at the front of the plane because the seats were in the aisle. [But] because the plane split open, we got a draft of air from the storm which kept smoke inhalation to a minimum. The rain did not put out the fire, but it slowed it down long enough for some of us to find the hole in the side of the plane. The smoke was real bad until we got the draft of fresh air. If the plane had not hit the runway lights, we would have landed in the river, and no one would have survived. If the plane had been any other model [of aircraft], it would have exploded, because the engine would have been in the wing. I consider myself very blessed to be here to fill out this form.